NOW 2 KNOW™

Algebra I

by T. G. D'Alberto

Pithy Professor Publishing Company
Brighton, CO

Published by

Pithy Professor Publishing Company, LLC
PO Box 33824
Northglenn, CO 80233

ISBN: 978-0-9882054-4-4

Library of Congress Control Number: 2014902795

Printed in the United States of America

About the Author

Dr. Tiffanie G. D'Alberto has a Ph.D. in Electrical & Computer Engineering from Cornell University and a B.S. and M.S. in Electrical Engineering from Virginia Polytechnic Institute & State University.

She has worked for over a decade in the telecommunications and aerospace industries as a scientist, program manager, and supervisor. She has engaged in numerous opportunities for tutoring, teaching, and mentoring throughout her career and schooling.

In her spare time, Tiffanie enjoys oil painting, drawing, reading, sewing, and running. She's a huge fan of Star Trek, Renaissance Festivals, and animals.

Tiffanie lives in St. Croix with her fiancé, Colin, and their many wonderful pets.

Dedication

To my dearest Colin, who inspires me, encourages me, and supports me. I could never thank you enough.

To my middle school Math teacher, Ms. Larkin, who provided the best foundation for learning math.

Acknowledgements

I always thank my family first: My parents for the foundation, the push, and the belief in me all along; My fiancé for his inspiration, encouragement, and unending support.

A huge thanks goes to my many excellent math teachers from middle school to high school to college that not only taught the material but also taught the way of thinking necessary to excel in these subjects.

I'd also like to acknowledge Barron's Review Course Series **Let's Review: Integrated Algebra** by Lawrence S. Leff, M.S. for a thorough refresher of the material. Any good book needs good graphics, and Softonic.com provided wonderful 3-D renditions for the graphing portion of the text.

Finally, I'd like to thank Amazon.com for their excellent publish-on-demand service that enables books such as these, and you, the reader, for making this investment in your future.

Table of Contents

Introduction

Welcome!

This course is unlike any you've taken thus far. The thinking required in Algebra 1 is very different from your other studies. Many students approach this course as though they are memorizing multiplication tables. Relax – you've already done that hard work. Algebra is easier than that. All you have to do is understand the methods and apply them.

The process of learning Algebra is three-fold:

1. **To excel at math is to understand math.** For example, you know how to play Go Fish. You not only know the rules, you understand the object of the game and the techniques that are required to dominate against your 4-year old opponent. It doesn't matter that this time you have a different hand. It doesn't matter that you haven't played in years. You *understand* the game, so you can play it well. That's how you should learn math.

2. **To understand math, you need the story** . The story is the logic flow that allows you to keep building on what you learn along the way. If someone tells you a story and skips a critical part of the plot, you would and should say, "Hey, back up!"

3. **To understand math, you also need the big picture.** The big picture is the outline of the story, placed in an area small enough for you to see it entirely. Like a file directory on a computer, it organizes the information. Once you see the flow of the big picture, it's easier for you to put the details of the story into their proper places.

The key to learning math is not memorization, it's *understanding*. Be open to changing the way you think. Once you get the flow, you'll get the A's. I wish you great success!

Layout:

The layout of this text is different from most academic books:

1. **The problem sets are saved to the end of the book.** In this book, you can read from beginning to end to understand the logical progression of the course, or stop to do problem sets as you desire.

2. **Solution sets give the critical steps to get the answers, not just the answers.** Because this is not a textbook for a classroom, there is no need to keep the "secret sauce" from you. Use the problems as drills or study the solutions as further examples.

3. **Appendix A is an overall summary of the entire book.** It helps you visualize the big picture and logic flow to give you a framework into which you can organize the details.

In addition, the following visual markers will help you navigate the material...

Key terms defined for the first time are **bolded** and also found in the index.

Important equations are shown as:

> *important information*

Illustrative graphics and additional notes are shown on the side to accompany the text.

Finally, examples are given as supplements to the text as well as for illustration:

> *Example* This is an example to illustrate a point or to give further definition. Skip it if you feel very comfortable with the material presented thus far.

Chapter 1: Numbers

Types of Numbers:

Let's start with three chapters of quick review. There are many types of numbers, some of which are outlined below:

Real Numbers: Real numbers include any number between $-\infty$ (negative infinity) and $+\infty$ (positive infinity). Examples include:
$$0, 1, 30, -2, \frac{1}{2}, \frac{13}{4}, 4.22222, -7.5146964099161601$$

Positive Numbers: Positive numbers are numbers that have a value greater than 0.

Negative Numbers: Negative numbers have a value less than 0. The real numbers contain positive numbers, negative numbers, and the number 0.

Integers or **Whole Numbers**: Integers are any real numbers with nothing after the decimal point. Examples include
$$-20, 0, 4, 365$$

Rational Numbers: Rational numbers can be expressed as an integer divided by a non-zero integer. They must either have a limited number of digits after the decimal or a repeating sequence of digits after the decimal. Examples include:
$$\frac{2}{1} = 2, \quad \frac{3}{4} = 0.75,$$
$$\frac{-50}{11} = -4.54545454\ldots = -4.\overline{54}$$

Irrational Numbers: Irrational numbers are numbers where the digits after the decimal go on forever without a pattern. Their decimal form can only be estimated (since no one can write digits out to infinity). Real numbers are either rational or irrational. Examples include:
$$22.565198129982\ldots, \ 3.141592654\ldots$$

Prime Numbers: A prime number is an integer that is greater than 1 and is NOT evenly divisible by integers other than itself and 1. Examples include:
$$2, 3, 5, 7, 11, 13, 17, 19$$

Composite Numbers: Composite numbers are integers greater than 1 that are not prime. Examples include:
$$4, 6, 8, 9, 10, 12, 14, 15, 16, 18, 20$$

Related to composite numbers are factors. A **factor** is an integer that is evenly divisible into another integer. As an example, all even numbers have a factor of 2.

Quick Factor Checks:

This is a factor	If this is evenly divided by it
2	Last digit
3	Sum of all digits
4	Last 2 digits
5	Last digit
8	Last 3 digits

Example

What kind of number is 25?

From the definitions, we can say 25 is:
  Real (on the number line from $-\infty$ to ∞)
  Positive (greater than 0)
  An Integer (no decimal places)
  Rational (can be expressed as $\frac{25}{1}$)
  A Composite (can be evenly divided by 5)

Example

What kind of number is -0.25?

From the definitions, we can say -0.25 is:
  Real (on the number line from $-\infty$ to ∞)
  Negative (less than 0)
  Rational (can be expressed as $\frac{-1}{4}$)

Example

What are factors of the number 20?

The number 20 is evenly divisible by 2, 4, 5, and 10.
So, 2, 4, 5, and 10 are factors of 20.

Working with Negative Numbers:

When working with negative numbers, the following table is helpful:

	+	−
+	+	−
−	−	+

In addition and subtraction, the table means:
a. Adding a positive number is adding
b. Adding a negative number is subtracting
c. Subtracting a positive number is subtracting
d. Subtracting a negative number is adding

	+	-
+	a	b
-	c	d

Reading the table for addition and subtraction.

Example

Evaluate:

$$3 + (+2) \qquad 3 + (-2) \qquad 3 - (+2) \qquad 3 - (-2)$$

Using the table above:

$$3 + 2 = 5 \qquad 3 - 2 = 1 \qquad 3 - 2 = 1 \qquad 3 + 2 = 5$$

For multiplication (and division), the table means:
a. A positive times a positive is a positive
b. A positive times a negative is a negative
c. A negative times a positive is a negative
d. A negative times a negative is a positive

	+	-
+	a	b
-	c	d

Reading the table for multiplication (division).

Example

Evaluate:

$$(+3) \times (+2) \qquad (+3) \times (-2) \qquad (-3) \times (+2) \qquad (-3) \times (-2)$$

Using the table above:

$$3 \times 2 = 6 \qquad 3(-2) = -6 \qquad -3(2) = -6 \qquad -3(-2) = 6$$

Note: Division is the same as multiplication except you multiply by a fraction:

$$3 \times \frac{1}{2} = \frac{3}{2} = 3 \div 2.$$

> **Example**
>
> Evaluate:
>
> $(+3) \div (+2)$ $(+3) \div (-2)$ $(-3) \div (+2)$ $(-3) \div (-2)$
>
> $3\left(\frac{1}{2}\right) = \frac{3}{2}$ $3\left(-\frac{1}{2}\right) = -\frac{3}{2}$ $-3\left(\frac{1}{2}\right) = -\frac{3}{2}$ $-3\left(-\frac{1}{2}\right) = \frac{3}{2}$

Besides addition, subtraction, multiplication, and division, there is another operation that comes up when dealing with negative numbers. Taking the **absolute value** of a number means forcing it to be positive. The symbol for absolute value is vertical lines on either side of the number.

> **Example**
>
> Take the absolute value of 2,-2, 45, and -4.53432
>
> Using the notation and definition:
>
> $|2| = 2;$
>
> $|-2| = 2;$
>
> $|45| = 45;$
>
> $|-4.53432| = 4.53432.$

Chapter 2: Fractions

Number Fractions:

Fractions are another way of showing division. Just like multiplication can be written with different symbols:

$$2 \times 3, \ 2 \cdot 3, \ 2 * 3, \text{ or } 2(3),$$

so can division:

$$2 \div 3, \ 2/3, \ \frac{2}{3}, \ 2\left(\frac{1}{3}\right).$$

In general, you can multiply or divide the top and bottom of a fraction by the same number. This is equivalent to multiplying by 1. As an example, if we multiply top and bottom of 2/3 by 4, we should still get 2/3:

$$\frac{2 \times 4}{3 \times 4} = \frac{2}{3} \times \frac{4}{4} = \frac{2}{3} \times 1 = \frac{2}{3}.$$

This method of "multiplying by 1" is very handy in making fractions easier to work with.

Reduction: To **reduce** a fraction, you simplify it by dividing **numerator** (top) and **denominator** (bottom) by all common factors.

Example

Reduce the fraction $\frac{18}{36}$:

$$\frac{18}{36} = \frac{18 \div 2}{36 \div 2} = \frac{9}{18} = \frac{9 \div 3}{18 \div 3} = \frac{3}{6} = \frac{3 \div 3}{6 \div 3} = \frac{1}{2}$$

Rationalization: When you have a **radical** (root) in the denominator, you can **rationalize** (get rid of it) by multiplying numerator and denominator by that radical.

Example

Rationalize $\frac{5}{\sqrt{2}}$.

$$\frac{5}{\sqrt{2}} = \frac{5 \times \sqrt{2}}{\sqrt{2} \times \sqrt{2}} = \frac{5\sqrt{2}}{2}$$

Addition & Subtraction: To add (or subtract) fractions, follow these steps:
- Find the **least common denominator** (LCD) – the smallest number that is evenly divisible by both denominators
- Multiply top and bottom by the same number to get the common denominator on one fraction
- Repeat the second step for the second fraction
- Add (or subtract) the numerators
- Reduce or rationalize as needed

Example

Evaluate $\frac{2}{3} + \frac{3}{4}$:

$$\frac{2}{3} = \frac{2 \times 4}{3 \times 4} = \frac{8}{12};$$

$$\frac{3}{4} = \frac{3 \times 3}{4 \times 3} = \frac{9}{12};$$

$$\frac{8}{12} + \frac{9}{12} = \frac{8+9}{12} = \frac{17}{12}$$

Multiplication & Division: To multiply (or divide) fractions, follow these steps:
- Multiply (or divide) the numerators
- Multiply (or divide) the denominators
- Reduce or rationalize as needed

Example

Evaluate $\frac{2}{3}\left(\frac{3}{4}\right)$:

$$\frac{2}{3}\left(\frac{3}{4}\right) = \frac{2 \times 3}{3 \times 4} = \frac{6}{12} = \frac{1}{2}$$

Real World Fractions:

In the practical world, a fraction with unitless numbers (i.e. just numbers) is used to describe a ratio, proportion, or percentage. A **ratio** and a **proportion** compare one thing to another. A **percentage** is a fraction of something expressed in 100th's.

> *Example*
>
> Ratio: There are 3 red jerseys for every 2 blue jerseys out on that field. Translation: Since there are a total of 5 jerseys in the group, $\frac{3}{5}$ of them are red, and $\frac{2}{5}$ of them are blue.
>
> Proportion: Four out of five dentists recommend XYZ toothpaste. Translation: XYZ toothpaste is recommended by $\frac{4}{5}$ of dentists.
>
> Percentage: That guy ate 40% of the pie! Translation: That guy ate $\frac{40}{100} = \frac{2}{5}$ of the pie.

When units get thrown in the mix, we can convert some quantities into others like distance and time into speed:

> *Example*
>
> A car takes 2 hours to cover 40 miles. How fast was the car was traveling?
>
> $\frac{40\ miles}{2\ hours} = \frac{40}{2}\ miles\ per\ hour = 20\ mph.$

We can also do unit conversion by repeatedly multiplying by quantities that are equivalent to 1:

> *Example*
>
> How many tablespoons are in 2 gallons?
>
> $\frac{2\ gallons}{}\frac{4\ quarts}{1\ gallon}\frac{2\ pints}{1\ quart}\frac{2\ cups}{1\ pint}\frac{16\ tbsp}{1\ cup} = 512\ tbsp$

Chapter 3: Exponents & Operations

Exponents:

An **exponent** is a way of getting a number (or **base**) to operate on itself. It is also called raising a number to a **power**, and it is denoted by a superscript. There are several types of exponents:

Positive Integer Exponents: Integer exponents greater than 0 indicate a number should be multiplied by itself exponent times:

$$2^4 = 2 \times 2 \times 2 \times 2 = 16; \quad 4^1 = 4$$

Negative Integer Exponents: Integer exponents less than 0 indicate a number should be multiplied by itself exponent times then inverted:

$$2^{-4} = \frac{1}{2 \times 2 \times 2 \times 2} = \frac{1}{16}; \quad 4^{-1} = \frac{1}{4}$$

"0" Exponent: Anything to the 0 power is 1:

$$2^0 = 1; \quad -4^0 = 1; \quad 349837.25^0 = 1$$

Fractional Exponents: Fractional exponents have two parts:
- The numerator is the number of times to multiply a number by itself;
- The denominator indicates taking a root of the number:

$$16^{1/4} = \sqrt[4]{16} = 2; \quad 4^{3/2} = \sqrt{4 \times 4 \times 4} = \sqrt{64} = 8$$

Negative Fractional Exponents: These work just like fractional exponents except the answer gets inverted:

$$16^{-1/4} = \frac{1}{\sqrt[4]{16}} = \frac{1}{2}; \quad 4^{-3/2} = \frac{1}{\sqrt{4 \times 4 \times 4}} = \frac{1}{\sqrt{64}} = \frac{1}{8}$$

Working with Exponents:

All of the operators we learned thus far can be applied to numbers with exponents.

Addition & Subtraction: There is no way to simplify addition or subtraction with numbers with exponents. Perform each exponent operation, then add or subtract.

$$2^2 - 3^2 = (2 \times 2) - (3 \times 3) = 4 - 9 = -5$$

Multiplication & Division: If the bases are the same,
- Multiplication is achieved by adding exponents;
- Division is achieved by subtracting exponents

$$2^2 \times 2^4 = 2^{2+4} = 2^6 = 64$$

Check: $(2 \times 2)(2 \times 2 \times 2 \times 2) = 4(16) = 64$

$$2^2 \div 2^4 = 2^{2-4} = 2^{-2} = \tfrac{1}{4}$$

Check: $\dfrac{(2 \times 2)}{(2 \times 2 \times 2 \times 2)} = \dfrac{4}{16} = \dfrac{1}{4}$

Raising to a Power: A number with an exponent can have another exponent applied to it by multiplying the two exponents:

$$(2^2)^3 = 2^{2 \times 3} = 2^6 = 64$$

Check: $(2 \times 2)(2 \times 2)(2 \times 2) = 4 \times 4 \times 4 = 64$

Absolute Value: An absolute value applied to a number with an exponent applies to the base only:

$$|(-2)^{-4}| = |-2|^{-4} = 2^{-4} = \frac{1}{2 \times 2 \times 2 \times 2} = \frac{1}{16}$$

Order of Operations & Properties:

The **order of operations** when evaluating an expression is:
- Parentheses & Absolute Values
- Exponents
- Multiplication & Division
- Addition & Subtraction

Note: Parentheses are implied with multiple elements in the numerator and denominator in division:
$\frac{2+4}{3+2} = (2 + 4) \div (3 + 2)$

Example

Evaluate $4 + 3(2 + 4) - 5(|3 - 2|)$

$4 + 3(2 + 4) - 5(|3 - 2|)$
$= 4 + 3(6) - 5(|-1|)$
$= 4 + 3(6) - 5(1)$
$= 4 + 18 - 5 = 17$

The **properties of addition and multiplication** are:

Commutative: Order doesn't matter:
$$2 + 3 = 3 + 2; \quad 2 \times 3 = 3 \times 2.$$

Associative: Grouping doesn't matter:
$$2 + (1 + 3) = (2 + 1) + 3 = (2 + 3) + 1$$
$$2(1 \times 3) = (2 \times 1)3 = (2 \times 3)1.$$

Distributive: Multiplication can be applied through an addition (or subtraction) in parentheses:
$$2 \times (4 + 3) = 2 \times 4 + 2 \times 3$$

Identity: The identity of a number turns it into itself.
- The **addition identity** is 0
$$2 + 0 = 2$$
- The **multiplication identity** is 1
$$2 \times 1 = 2$$

Inverse: The inverse of a number turns it into an identity.
- The **additive inverse** is the negative of a number
$$2 + (-2) = 0$$
- The **multiplicative inverse** is the number to the (-1) power
$$2 \times 2^{-1} = \frac{2}{2} = 1$$

Chapter 4: Equations

Equations & Variables:

Now we can start getting into Algebra! An **equation** or an **equality** is a way to relate two things by saying they are equal. You've dealt with equations many times before:

$$1 + 2 = 3; \quad 2 \times 2 = 4$$

In Algebra, we write equations in a slightly different way Consider the following equation:

$$\boxdot + 2 = 3$$

What goes in the box? Intuitively, you would say the answer is 1. Similarly, if presented with this equation:

$$\boxdot \times 2 = 4$$

you would hopefully say that the number 2 should go in the box. The values that make these equations true (1 in the first problem, and 2 in the second) are called **solutions** to the equations.

In Algebra, instead of drawing a box, we use a letter. The letter x is very popular, but any of them could be used. We can rewrite the two equations above as:

$$x + 2 = 3; \quad x \times 2 = 4$$

The first equation has the solution $x = 1$, and the second has the solution $x = 2$.

Sometimes new students of Algebra get confused as to how x can be one value one time and equal to something entirely different the next time. That's because x is nothing more than an empty box. And, because its value can change from equation to equation, it is called a **variable**.

A Few More Definitions:

We have a few more terms to throw around so that everyone in the math world can speak the same language. Numbers in an equation never change meaning or value, so they are called **constants**. For example, the 2 and 3 are constants in the equation:

$$x + 2 = 3.$$

When multiplying a constant and a variable, instead of using an X, dot, star, or parentheses, it's okay to write the constant and variable next to each other without a space. For example, the equation $x \times 2 = 4$ is usually written:

$$2x = 4$$

The constant multiplying the variable (or powers of the variable) is called the **coefficient**. Above, the 2 is the coefficient of x.

To summarize the notation, consider the expression:

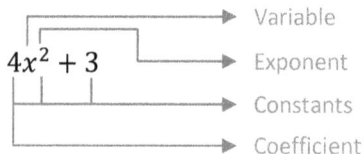

$$4x^2 + 3$$

- Variable
- Exponent
- Constants
- Coefficient

We can combine constants, variables, exponents, coefficients, and operators to make up **algebraic expressions**, like the one above. When we separate algebraic expressions with an equals sign, we have an equation.

Much of Algebra 1 is focused on finding solutions to equations with variables. This is a nice segue to our next section.

Solving Simple Equations:

Isaac Newton said - For every action, there is an equal and opposite reaction. So it is with Algebra.

Consider our current chapter examples. Other than relying on your memorization skills with addition and multiplication tables, there is a better way to solve those equations:

> **What you do to one side of an equation,**
> **you must also do to the other.**

Consider our first equation:

$$x + 2 = 3.$$

We want to isolate the variable on one side of the equation so that we get a solution that reads $x = \cdots$. To do this, we need to get rid of the 2. If we must subtract 2 from one side, then we must also subtract 2 from the other side.

$$(-2) + x + 2 = 3 + (-2)$$
$$x + (2 - 2) = 3 - 2$$
$$x = 1$$

To check a solution, plug it back into the original equation:
$1 + 2 = 3$

Now, consider our second equation:

$$2x = 4$$

Once again, we have a 2 to eliminate - this time by division:

$$\frac{1}{2} \times 2x = 4 \times \frac{1}{2}$$
$$x = 2$$

Check:
$2(2) = 4$

NOTE: You can divide both sides of an equation by a constant or variable, as long as you ***DO NOT DIVIDE BY 0***.

Now let's consider a more complicated example:
$$4x + 2 = 14$$

In this equation, if we want to get something that looks like $x = \cdots$, we have to get rid of a 4 and a 2. Let's subtract off the 2 first:
$$[4x + 2] - 2 = [14] - 2$$
$$4x = 12$$

Next, we will get rid of the 4 with division:

$$\tfrac{1}{4} \times [4x] = [12] \times \tfrac{1}{4}$$
$$x = \frac{12}{4} = 3$$

Check:
$$4(3) + 2 = 12 + 2 = 14$$

The solution is $x = 3$. Now, let's consider what happens if we start with the original equation and this time get rid of the 4 first:

$$\tfrac{1}{4} \times [4x + 2] = [14] \times \tfrac{1}{4}$$
$$\frac{4x}{4} + \frac{2}{4} = \frac{14}{4}$$
$$x + \frac{1}{2} = \frac{7}{2}$$

Notice that we divided by 4 across the ENTIRE equation, not just the $4x$ part. Well, things have gotten messy, but we are almost there. We have to get rid of a ½ by subtraction:

$$x + \frac{1}{2} - \frac{1}{2} = \frac{7}{2} - \frac{1}{2}$$
$$x = \frac{7-1}{2} = \frac{6}{2} = 3$$

So, we get the same answer, $x = 3$, but it was messier to do it this way. As a rule of thumb to make life easier:

**Addition & Subtraction first,
Multiplication & Division last.**

Let's do some more examples:

> **Example**
>
> Solve the equation for x:
>
> $$3x + 4 = 10$$
>
> $$[3x + 4] - 4 = [10] - 4$$
>
> $$3x = 6$$
>
> $$\frac{1}{3} \times 3x = 6 \times \frac{1}{3}$$
>
> Check:
>
> $$x = 2$$
>
> $3(2) + 4 = 6 + 4 = 10$

> **Example**
>
> Solve the equation for x:
>
> $$5x - 1 = 14$$
>
> $$[5x - 1] + 1 = [14] + 1$$
>
> $$5x = 15$$
>
> $$\frac{1}{5} \times 5x = 15 \times \frac{1}{5}$$
>
> Check:
>
> $$x = 3$$
>
> $5(3) - 1 = 15 - 1 = 14$

> **Example**
>
> Solve the equation for x:
>
> $$\frac{x}{3} - 1 = 2$$
>
> $$\left[\frac{x}{3} - 1\right] + 1 = [2] + 1$$
>
> $$\frac{x}{3} = 3$$
>
> $$3 \times \frac{x}{3} = 3 \times 3$$
>
> Check:
>
> $$x = 9$$
>
> $\frac{9}{3} - 1 = 3 - 1 = 2$

An Alternate Solving Technique:

There is another way to think about solving equations. Instead of performing an action directly on an equation, you can add, subtract, multiply, or divide by another equation that is true. Let's look at some of the prior examples using this technique:

Example

Solve the equation for x:

$$3x + 4 = 10$$
$$+ \quad -4 = -4 \qquad \leftarrow \text{Add a True equation}$$
$$\overline{3x \quad\quad = 6}$$

$$3x = 6$$
$$\div \quad 3 = 3 \qquad \leftarrow \text{Divide by a True equation}$$
$$\overline{x = \frac{6}{3}}$$

$$x = 2$$

Example

Solve the equation for x:

$$5x - 1 = 14$$
$$+ \quad\quad 1 = 1 \qquad \leftarrow \text{Add a True equation}$$
$$\overline{5x \quad\quad = 15}$$

$$5x = 15$$
$$\div \quad 5 = 5 \qquad \leftarrow \text{Divide by a True equation}$$
$$\overline{x = 3}$$

You can use whichever method you feel most comfortable with. In either case, you are doing the same thing – adding and subtracting constants, then multiplying and dividing by constants.

Chapter 5: More Complex Equations

Solving Equations with Variables with Exponents:

When dealing with variables with exponents, save the exponents for last. Just like when deciding to take care of addition and subtraction before multiplication and division, you will be avoiding a big mess by saving the exponents for after these operations. As a rule of thumb;

> **Reverse the typical order of operations to solve an equation:**
>
> **Addition & Subtraction**
> **Multiplication & Division**
> **Exponents**

Let's look at an example:

Example

Solve the equation for x:
$$4x^2 - 4 = 12$$

$$4x^2 - 4 + 4 = 12 + 4 \quad \longleftarrow \text{Add \& Subtract first}$$
$$4x^2 = 16$$

$$\frac{1}{4} \times 4x^2 = 16 \times \frac{1}{4} \quad \longleftarrow \text{Multiply \& Divide next}$$
$$x^2 = 4 \quad \longleftarrow \text{Then Exponents}$$

Once again, what you do to one side, you do to the other:

$$\sqrt{x^2} = \sqrt{4}$$

$$x = \pm 2$$

Check:
$$4(2)(2) - 4 = 4(4) - 4$$
$$= 16 - 4 = 12$$

$$4(-2)(-2) - 4 = 4(4) - 4$$
$$= 16 - 4 = 12$$

Notice in the example that we have two solutions: $x = 2$ and $x = -2$. That's because:

> **Even roots of a positive number can be negative or positive.**

Solving Equations with Parentheses:

So far, we've learned how to solve equations with addition, subtraction, multiplication, division, and exponents. What happens when you have algebraic expressions with parentheses involved? As an example:

$$4(x - 1) = 8$$

There are two ways to deal with parentheses:

> - **Simplify the expressions to get rid of the parentheses and then solve as usual, or**
> - **Save the parentheses for last.**

Let's solve the above equation both ways to illustrate. In the first method, we will multiply the 4 across the parentheses and then solve the equation:

$$4(x - 1) = 8$$

$$4x - 4 = 8 \qquad \longleftarrow \text{Get Rid of Parentheses}$$

$$4x - 4 + 4 = 8 + 4 \qquad \longleftarrow \text{Solve as Usual}$$

$$4x = 12$$

$$\frac{1}{4} \times 4x = 12 \times \frac{1}{4}$$

$$x = 3 \qquad\qquad \text{Check:}$$
$$4(3 - 1) = 4(2) = 8$$

Now we'll save the parentheses for last to illustrate the second solving method:

$$4(x - 1) = 8$$

$$\frac{1}{4} \times [4(x - 1)] = 8 \times \frac{1}{4} \qquad \longleftarrow \text{Start Solving as Usual}$$

$$(x - 1) = 2 \qquad \longleftarrow \text{( ) No Longer Needed}$$

$$x - 1 + 1 = 2 + 1$$

$$x = 3$$

Here are some more examples with the equations solved both ways:

Example

Solve the equation for x: $2(x - 3) = 4$

Method 1	Method 2
$2(x - 3) = 4$	$2(x - 3) = 4$
$2x - 6 = 4$	$\frac{1}{2} \times 2(x - 3) = 4 \times \frac{1}{2}$
$2x - 6 + 6 = 4 + 6$	$(x - 3) = 2$
$2x = 10$	$x - 3 + 3 = 2 + 3$
$\frac{1}{2} \times 2x = 10 \times \frac{1}{2}$	$x = 5$
$x = 5$	Check: $2(5 - 3) = 2(2) = 4$

Example

Solve the equation for x: $3(2x - 3) + 6 = -9$

Method 1	Method 2
$3(2x - 3) + 6 = -9$	$3(2x - 3) + 6 = -9$
$6x - 9 + 6 = -9$	$\frac{1}{3} \times [3(2x - 3) + 6] = -9 \times \frac{1}{3}$
$6x - 3 = -9$	$(2x - 3) + 2 = -3$
$6x - 3 + 3 = -9 + 3$	$2x - 1 = -3$
$6x = -6$	$2x - 1 + 1 = -3 + 1$
$\frac{1}{6} \times 6x = -6 \times \frac{1}{6}$	$2x = -2$
$x = -1$	$\frac{1}{2} \times 2x = -2 \times \frac{1}{2}$
	$x = -1$

Check:
$$3((2(-1) - 3) + 6$$
$$= 3(-2 - 3) + 6$$
$$= 3(-5) + 6$$
$$= -15 + 6 = -9$$

When the parentheses are replaced by absolute value signs, the problem gets split in two:

Example

Solve the equation for x: $2|x - 3| = 4$

$$2|x - 3| = 4$$

$$|x - 3| = 2$$

$$x - 3 = \pm 2$$

Getting rid of the absolute value sign splits the problem into two equations:

$$x - 3 = 2 \qquad x - 3 = -2$$

$$x = 5 \qquad\quad x = 1$$

Check:
$2|5 - 3| = 2(2) = 4$
$2|1 - 3| = 2(2) = 4$

Example

Solve the equation for x: $3|2x - 3| + 6 = 9$

$$3|2x - 3| + 6 = 9$$

$$3|2x - 3| = 3$$

$$|2x - 3| = 1$$

$$2x - 3 = \pm 1$$

Getting rid of the absolute value sign splits the problem into two equations:

$$2x - 3 = 1 \qquad 2x - 3 = -1$$

$$2x = 4 \qquad\quad 2x = 2$$

$$x = 2 \qquad\quad x = 1$$

Check:
$3|2(2) - 3| + 6$
$= 3|1| + 6 = 9$

$3|2(1) - 3| + 6$
$= 3|-1| + 6 = 9$

Chapter 6: Isolation & Grouping

Solving Equations with Variables on Both Sides:

Sometimes you get a problem where the variable happens to be on both sides of the equation. But, you want a solution that looks like

$$x = \cdots$$

To achieve the above form, you need to **isolate** the variable, i.e. put the variable on one side of the equation by itself.

You've been doing this all along, and you'll use the same techniques as before. The only difference is that you need to know how to add and subtract with variables.   This is done by **grouping like terms** - taking terms that are like each other and putting them together. To use math language, we will be using the distributive property.

Let's evaluate the following expression:
$$2x + 5x$$

This is the same as writing:
$$x(2 + 5) = x(7) = 7x$$

We grouped the terms and simplified the expression. To further illustrate the point, consider this more complicated example:
$$3x - 5 - x + 2 + 4(x - 2)$$

First get rid of the parentheses:
$$3x - 5 - x + 2 + 4x - 8$$

Now group like terms – variable terms versus constants:
$$(3x - x + 4x) + (-5 + 2 - 8)$$
$$= x(3 - 1 + 4) + (-11)$$
$$= 6x - 11$$

Let's use isolation and grouping like terms on equations.

Example

Solve the equation for x:
$$3x + 1 = x - 3$$

We start by subtracting x from both sides, then proceed as usual:
$$3x + 1 - x = x - 3 - x$$

Grouping like terms:
$$x(3 - 1) + 1 = x(1 - 1) - 3$$
$$2x + 1 = -3$$
$$2x + 1 - 1 = -3 - 1$$
$$2x = -4$$
$$\frac{1}{2} \times 2x = -4 \times \frac{1}{2}$$
$$x = -2$$

Check:
$$3(-2) + 1 = -6 + 1 = -5$$
$$-2 - 3 = -5$$
Both sides agree.

Example

Solve the equation for x:
$$x + 4 = -2x - 1$$

We start by adding $2x$ to both sides, then proceed as usual:
$$x + 4 + 2x = -2x - 1 + 2x$$

Grouping like terms:
$$x(1 + 2) + 4 = x(-2 + 2) - 1$$
$$3x + 4 = -1$$
$$3x + 4 - 4 = -1 - 4$$
$$3x = -5$$
$$\frac{1}{3} \times 3x = -5 \times \frac{1}{3}$$
$$x = -\frac{5}{3}$$

Check:
$$-\frac{5}{3} + 4 = \frac{-5 + 12}{3} = \frac{7}{3}$$
$$-2\left(-\frac{5}{3}\right) - 1 = \frac{10}{3} - 1$$
$$= \frac{(10 - 3)}{3} = \frac{7}{3}$$
Both sides agree.

Preview to Factoring:

One final note on grouping like terms: Sometimes you may end up with a situation where it looks like you can divide the whole equation by a variable to simplify things. You can do this as long as you make a note that $x = 0$ is one of the solutions. You are not really dividing by 0. You are just breaking the problem into two parts by factoring. This will be explained in more detail in Chapter 10.

Example

Solve the equation for x:
$$3x^2 + x = -x^2 + 2x$$

We start by adding x^2 to both sides:
$$3x^2 + x^2 + x = -x^2 + 2x + x^2$$
$$4x^2 + x = 2x$$
Next we subtract $2x$ from both sides:
$$4x^2 + x - 2x = 2x - 2x$$
$$4x^2 - x = 0$$

Make a note that $x = 0$ is one of the solutions, and get rid of the extra x:

$$\frac{1}{x} \times [4x^2 - x] = 0 \times \frac{1}{x}$$

Check:
$$3(0^2) + 0 = -(0^2) + 2(0)$$

Solve as usual:
$$4x - 1 = 0$$
$$4x - 1 + 1 = 0 + 1$$
$$\frac{1}{4} \times 4x = 1 \times \frac{1}{4}$$
$$x = \frac{1}{4}$$

The solutions are $x = 0$, and $x = \frac{1}{4}$.

Check:
$$3\left(\frac{1}{4}\right)\left(\frac{1}{4}\right) + \frac{1}{4} = \frac{3}{16} + \frac{1}{4}$$
$$= \frac{3+4}{16} = \frac{7}{16}$$

$$-\left(\frac{1}{4}\right)\left(\frac{1}{4}\right) + 2\left(\frac{1}{4}\right) = \frac{-1}{16} + \frac{2}{4}$$
$$= \frac{-1+8}{16} = \frac{7}{16}$$
Both sides agree.

Chapter 7: Sets

Set Notation:

You might not have formally learned about sets, but you have used them often. A **set** is a collection of things: toys, cars, music, or numbers. When we have more than one solution to an equation, we'll talk about a set of solutions.

There are several ways to write sets. The first is **roster form** – listing each **element** (member) of the set between squiggly parentheses. As an example, let's say we have a set, call it A, of the numbers 2, 3, 5, and 8. If we want to say that a number, call it x, is part of this set, we write:
$$A = \{2,3,5,8\} \text{ and } x \in A$$

The above reads, "A is the set of numbers 2, 3, 5, and 8 and x is an element of the set A." It means that x can take on any of those values.

Now let's say that x is a number *greater than or equal to* 1, but *less than* 2, i.e. $1 \leq x < 2$. That means x could be 1, 1.1, 4/3, 1.555, 1.999999, etc. Obviously, the possible values would be too numerous to list, so we could use **interval notation**:
$$x \in [1,2)$$

"Interval notation" isn't formal set notation, but it is a nice shorthand.

The closed parenthesis, [, on the 1 indicates that x could be equal to 1. The open parenthesis,), on the 2 indicates that x can be almost, but not equal to 2. **Graphically**, we can also show the set as it appears to the right. Circles indicate endpoints; an open circle means up to but not including.

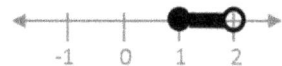

Graphical representation of $A = \{x : 1 \leq x < 2\}$

Finally, we could also use **set-builder notation**. For the set above, this would look like:
$$A = \{x : 1 \leq x < 2\} \text{ or } A = \{x \mid 1 \leq x < 2\}$$

The colon and vertical line are interchangeable. The above reads, "A is a set of numbers x such that $1 \leq x < 2$." Any additional rules would be put inside the squiggly parentheses separated by semi-colons.

Special Sets:

The biggest number set you've encountered in this book is the set of real numbers. This set consists of all values from $-\infty$ to $+\infty$. No number can truly equal infinity, so we use open parentheses to write:

$$A = \{x : x \in (-\infty, \infty)\} \text{ or } x \in A = \{x : -\infty < x < \infty\}$$

The set of real numbers is so useful, it gets a special notation, $\mathbb{R}$. The above could also be written:
$$A = \{x : x \in \mathbb{R}\}.$$

Special sets include:

> $\mathbb{R}$ - **The real numbers**
> $\mathbb{N}$ - **All positive integers: {1, 2, 3, ...}**
> $\mathbb{Z}$ - **All integers: {..., -2, -1, 0, 1, 2, ...}**
> $\mathbb{Q}$ - **Rational numbers: {- ½, ¾, ...}**
> $\emptyset$ - **The null set: an empty set: { }, not {0}**

There are a couple of more terms related to sets. A **subset** is a set within a set. A **superset** is a set that contains a set. For example, the set of all integers, $\mathbb{Z}$, is a subset of the real numbers, $\mathbb{R}$. In math parlance, we write this as:

$$\mathbb{Z} \subseteq \mathbb{R}$$

Or, you could say that the real numbers are a superset that contain the set of all integers. In math parlance, this is written as:
$$\mathbb{R} \supseteq \mathbb{Z}$$

A subset is a set within a set. A superset is a set that holds another set.

Now that we have a working vocabulary for sets, we can do some work with them.

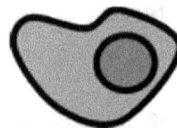

Working With Sets:

There are three things you can do with sets right away:
Find the complement, the union, and the intersection.

Complement: If we have a set, and if A is a subset within that set, the complement A is everything in the set that isn't in A. The complement of A is written as A'.

Example
> Find the complement of
> $$A = \{x: x \leq 1; x \in \mathbb{Z}\}$$
>
> The above states that A is the set of values less than or equal to 1 that are also integers. In other words, we have:
> > The set: all integers
> > The subset, A: all integers with $x \leq 1$
>
> So, the complement would be all integers with $x > 1$ which we write as:
> > $A' = \{x: x > 1; x \in \mathbb{Z}\}$

Union: If we have two sets, A and B, then the set of elements that are in A, B, or both is called the union of A and B. We write A union B as:

$$A \cup B$$

Intersection: If we have two sets, A and B, then the set of elements that are in both A and B is called the intersection of A and B. We write A intersect B as:

$$A \cap B$$

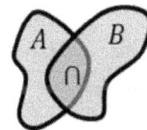

The union is everything pictured; the intersection is where they overlap.

The next page shows examples of unions and intersections.

Find the union and intersection of

$$A = \{1, 3, 5, 7\}$$
$$B = \{2, 5, 10\}$$

Example

Union: $A \cup B = \{1, 2, 3, 5, 7, 10\}$

Union: A and/or B

Intersection: $A \cap B = \{5\}$

Intersection: A and B

Find the union and intersection of

$$A = \{-1, 0, 6, 8\}$$
$$B = \{-1, 2, 6, 8\}$$

Example

Union: $A \cup B = \{-1, 0, 2, 6, 8\}$

Union: A and/or B

Intersection: $A \cap B = \{-1, 6, 8\}$

Intersection: A and B

Find the union and intersection of

$$A = \{x : 1 < x < 5; x \in \mathbb{Z}\}$$

$A = \{2, 3, 4\}$

Example

$$B = \{-2, -\tfrac{4}{5}, 0, \tfrac{1}{2}\}$$

Union: $A \cup B = \{-2, -\tfrac{4}{5}, 0, \tfrac{1}{2}, 2, 3, 4\}$

Union: A and/or B

Intersection: $A \cap B = \emptyset$
(there is nothing that appears in both sets)

Intersection: A and B

Chapter 8: Inequalities

Introduction to Inequalities:

In your work solving equations thus far, you found very specific solutions. In other words, there were one or two values for which the variable worked in the equality.

In the case of an **inequality**, a relation without an equals sign, you get a set of solutions. As an example:

$$x - 3 > 5$$

Following the same rules you used with equalities, we can add 3 to both sides:

$$x - 3 + 3 > 5 + 3$$
$$x > 8$$

Solving the inequality gave us an infinite number of solutions: $\{8.5, 9, 10, 22.22, ...\}$. We can write the solution set as: $S = \{x : x > 8\}$ or $x \in (8, \infty)$.

We can also end up with a problem that has two inequalities in the relation:

$$1 < x - 5 \leq 2$$

We want to get rid of the 5, so we add 5 to ALL parts of the relation:

$$1 + 5 < x - 5 + 5 \leq 2 + 5$$
$$6 < x \leq 7$$

The solution set is now $S = \{x : 6 < x \leq 7\}$ or $x \in (6,7]$.

Inequalities with Multiplication & Division:

As you've seen, you can add or subtract in inequalities just as you did with equalities - just be sure to treat all sides of any relation. You can also use isolation and grouping of like terms. Multiplication and division are a little different, though.

Let's consider the inequality:

$$2 > 1$$

This is a true statement. If we multiply both sides by 2, we get:

$$4 > 2$$

which is still a true statement. What if, instead, we had multiplied by -2? Then we would have:

$$-4 > -2$$

This is NOT a true statement. This leads to a very important rule when working with inequalities:

> **When multiplying or dividing by a negative number, you must reverse the inequality.**

The correct way to multiply through by -2 is:

$$2 > 1$$
$$2 \times (-2) < 1 \times (-2)$$
$$-4 < -2.$$

Let's work some examples of solving inequalities.

Example → Solve: $2x - 4 > 2$

$$2x - 4 + 4 > 2 + 4$$

$$2x > 8$$

$$\frac{1}{2} \times 2x > 8 \times \frac{1}{2}$$ ½ > 0, so sign stays the same

$$x > 4$$

$$S = \{x: x > 4; x \in \mathbb{R}\}$$

Example → Solve: $2x - 2 \geq 4x + 2$

$$2x - 2 + 2 \geq 4x + 2 + 2$$

$$2x \geq 4x + 4$$

$$2x - 4x \geq 4x + 4 - 4x$$

$$-2x \geq 4$$

$$\left(\frac{-1}{2}\right) \times (-2x) \leq 4 \times \left(\frac{-1}{2}\right)$$ -½ < 0, so sign reverses

$$x \leq -2$$

$$S = \{x: x \leq 2; x \in \mathbb{R}\}$$

Example → Solve: $0 < 3 - 4x \leq 9$

$$0 - 3 < 3 - 4x - 3 \leq 9 - 3$$

$$-3 < -4x \leq 6$$

$$\left(\frac{-1}{4}\right) \times (-3) > \left(\frac{-1}{4}\right) \times (-4x) \geq 6 \times \left(\frac{-1}{4}\right)$$ -¼ < 0, so ALL signs reverse

$$\frac{3}{4} > x \geq -\frac{6}{4}$$

$$\frac{3}{4} > x \geq -\frac{3}{2}$$

$$S = \{x: \frac{3}{4} > x \geq -\frac{3}{2}; x \in \mathbb{R}\}$$

Inequalities with Absolute Values:

The absolute value sign is another tricky operation with inequalities. Just as with equalities, it gives us two problems to solve. The steps are:

> - **Isolate the absolute value to the left side of the equation.**
> - **To set up the first problem, we drop the absolute value sign, solve the equation as is.**
> - **To set up the second problem, we drop the absolute value, reverse the inequality, and negate the right hand side.**

Example

Solve: $|2 - x| > 4$

The absolute value sign is already isolated to the left side of the inequality. We set up the two problems as follows:

$$2 - x > 4 \qquad\qquad 2 - x < -4$$
$$-x > 2 \qquad\qquad -x < -6$$
$$x < -2 \qquad\qquad x > 6$$

$$S = \{x : x < -2; x > 6; x \in \mathbb{R}\}$$

Note the multiplication of both sides by -1 forces an inequality reversal in both problems

Check:
  Since $-3 < -2$, we check the first equation:
  $$|2 - (-3)| = |5| = 5 > 4 \ \text{ ok}$$

  Since $7 > 6$, we check the second equation:
  $$|2 - 7| = |-5| = 5 > 4 \ \text{ ok}$$

Chapter 9: Polynomials

Definitions:

A **polynomial** is an algebraic expression with one or more unlike terms added or subtracted together. Some polynomials are given special names:

- **Monomials** have one term:
    For example: $2x$, 4, 10

- **Binomials** have two terms:
    For example: $2x + 4$, $5x - 9$

- **Trinomials** have three terms:
    For example: $x + y + 2$, $2x + 3y - 7$

Up until now, you've looked at a number of monomials and binomials, but algebraic expressions can have many unlike terms. Examples of unlike terms include:
- Constants vs. variables
- Different variables
- Different powers of the same variable

The last item, different powers of the same variable, provides another way to describe a polynomial. The highest power in the expression is the **degree** of a polynomial.
    For example, the polynomial $5x^4 + 2x$
    is a fourth degree binomial in x.

A polynomial is written in **standard form** when all of the elements are in order of decreasing power.
    For example: $2x^5 - 4x^3 + x - 2.$

Note: Since a constant can be written as:

$$constant \times x^0 = constant \times 1 = constant,$$

the constant will go between terms with x and $1/x$ (or x^1 and x^{-1}).

Adding & Subtracting Polynomials:

Adding and subtracting within a polynomial just means grouping like terms.

> *Example*
>
> Simplify the following polynomial:
> $$3x^2 + 5y - 2x^2 + 2x - 4y + 2$$
>
> $$(3 - 2)x^2 + 2x + (5 - 4)y + 2$$
> $$x^2 + 2x + y + 2$$

Adding two or more polynomials is basically the same thing. You throw all of their terms together and simplify. If you line them on top of each other, it makes the bookkeeping easier:

> *Example*
>
> Add the following polynomials:
> $$3x^2 + 4x + 5; \quad x^2 - 2$$
>
> Arranging them and summing looks like this:
>
> $$\begin{array}{r} 3x^2 + 4x + 5 \\ + \ x^2 \qquad - 2 \\ \hline 4x^2 + 4x + 3 \end{array}$$

To subtract one from the other, you multiply the one being subtracted by (-1) and then sum the two expressions:

> *Example*
>
> Subtract the second polynomial from the first:
> $$3x^2 + 4x + 5; \quad x^2 - 2$$
>
> First change the one being subtracted, then add:
>
> $$(-1) \times [x^2 - 2] = -x^2 + 2$$
>
> $$\begin{array}{r} 3x^2 + 4x + 5 \\ + \ -x^2 \qquad +2 \\ \hline 2x^2 + 4x + 7 \end{array}$$
>
> This is the same as doing straight subtraction. Just be sure to SUBTRACT all of the terms of the 2nd expression:
>
> $$\begin{array}{r} 3x^2 + 4x + 5 \\ - \ [x^2 \qquad - 2] \\ \hline 2x^2 + 4x + 7 \end{array}$$

Multiplying Polynomials:

Do you remember learning how to multiply multi-digit numbers together? Multiplying polynomials is the same thing:

> - **Stack one on top of the other**
> - **Create each line as a term of the bottom polynomial times all of the terms of the top polynomial**
> - **Add all the lines to get the final expression.**

You'll find that if you align the like terms in columns, it will make the bookkeeping easier. Let's look at some examples:

Example

Multiply $x^2 - 3$ and $3x^2 + 2x - 2$.

The first term is shorter, so we'll put it on the bottom:

$$
\begin{array}{r}
3x^2 + 2x - 2 \\
\times \quad x^2 - 3 \\
\hline
-9x^2 - 6x + 6 \quad \leftarrow \text{Multiply by } -3 \\
+\ 3x^4 + 2x^3 - 2x^2 \qquad\quad \leftarrow \text{Multiply by } x^2 \\
\hline
3x^4 + 2x^3 - 11x^2 - 6x + 6 \quad \leftarrow \text{Add it all together}
\end{array}
$$

Notice that multiplying variables with exponents is just like with constants: Add the exponents of like bases.

Example

Multiply $x^2 - x + 2$ and $2x^2 + x - 1$.

$$
\begin{array}{r}
x^2 - x + 2 \\
\times \quad 2x^2 + x - 1 \\
\hline
-x^2 + x - 2 \quad \leftarrow \text{Multiply by } -1 \\
x^3 - x^2 + 2x \qquad\quad \leftarrow \text{Multiply by } x \\
+\ 2x^4 - 2x^3 + 4x^2 \qquad\qquad \leftarrow \text{Multiply by } 2x^2 \\
\hline
2x^4 - x^3 + 2x^2 + 3x - 2 \quad \leftarrow \text{Add it all together}
\end{array}
$$

The above examples show how to multiply any two polynomials. To multiply two binomials, there is a handy rule called FOIL:

FOIL = First + Outer + Inner + Last

FOIL is the way to remember to multiply all of the terms:

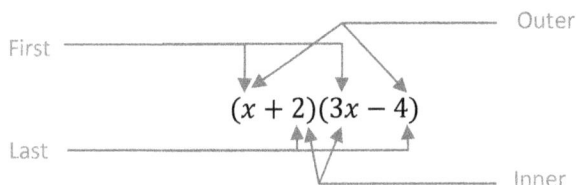

The solution is:

First: $x \times 3x = 3x^2$
Outer: $x \times -4 = -4x$
Inner: $2 \times 3x = 6x$
Last: $2 \times -4 = -8$

Adding it together gives:
$$3x^2 - 4x + 6x - 8 = 3x^2 + 2x - 8$$

You could employ the stacking method, but the FOIL method for binomials is quick and it gives insight into factoring which will be discussed in the next chapter.

Example

Evaluate $(x + 1)(2x - 2)$.

$\quad$ F $\quad$ + $\quad$ O $\quad$ + $\quad$ I $\quad$ + $\quad$ L
$2x^2 - 2x + 2x - 2 = 2x^2 - 2$

Example

Evaluate $(2x + 3)(x - 5)$.

$\quad$ F $\quad$ + $\quad$ O $\quad$ + $\quad$ I $\quad$ + $\quad$ L
$2x^2 - 10x + 3x - 15 = 2x^2 - 7x - 15$

Polynomial Fractions:

You can treat fractions with polynomials in the numerator and denominator just like you do fractions with numbers in the numerator and denominator.

> - **To multiply fractions, multiply numerators and multiply denominators**
> - **To divide fractions, invert the divisor, then multiply numerators and multiply denominators**
> - **To add or subtract fractions, find a common denominator, modify each fraction to have that denominator, then perform the indicated addition or subtraction**

Example

Evaluate $\frac{x+2}{x-1}\left(\frac{x+1}{x-1}\right)$.

$$\frac{x+2}{x-1}\left(\frac{x+1}{x-1}\right) = \frac{(x+2)(x+1)}{(x-1)(x-1)} = \frac{x^2+2x+x+2}{x^2-x-x+1} = \frac{x^2+3x+2}{x^2-2x+1}$$

Example

Evaluate $\frac{x+2}{x-1} \div \left(\frac{x+1}{x-1}\right)$.

$$\frac{x+2}{x-1} \div \left(\frac{x+1}{x-1}\right) = \frac{x+2}{x-1}\left(\frac{x-1}{x+1}\right)$$

The $(x-1)$ terms cancel:

$$\frac{x+2}{\cancel{x-1}}\left(\frac{\cancel{x-1}}{x+1}\right) = \frac{(x+2)}{(x+1)}.$$

Example

Evaluate $\frac{x+2}{x-1} + \left(\frac{x}{x-3}\right)$.

$$\frac{x+2}{x-1} + \left(\frac{x}{x-3}\right) = \frac{(x+2)(x-3)}{(x-1)(x-3)} + \frac{x(x-1)}{(x-1)(x-3)}$$

$$= \frac{(x^2-3x+2x-5)+(x^2-x)}{x^2-3x-x+3} = \frac{2x^2-2x-5}{x^2-4x+3}$$

Dividing Polynomials:

Dividing two polynomials is just like doing long division:

> • **Place one outside the division sign (divisor), the other inside (dividend)**
> • **Put multipliers on top (quotients)**
> • **Subtract each multiplied expression from the dividend**
> • **Drop the next term down**
> • **Express the remainder as a fraction of the divisor.**

Let's look at some examples:

Example

Divide x into $3x^2 + 2x - 2$.

$$
\begin{array}{r}
3x + 2 \\
x \overline{)3x^2 + 2x - 2} \\
-3x^2 \qquad\qquad \\
\overline{2x} \\
- 2x \\
\overline{-2}
\end{array}
$$

        $3x$ component

        $+2$ component

        Remainder

The answer is $3x + 2 - \dfrac{2}{x}$

Example

Divide $x + 1$ into $2x^2 + 4x + 3$.

$$
\begin{array}{r}
2x + 2 \\
x + 1 \overline{)2x^2 + 4x + 3} \\
-[2x^2 + 2x] \qquad\quad \\
\overline{2x + 3} \\
-[2x + 2] \\
\overline{1}
\end{array}
$$

        $2x$ component

        $+2$ component

        Remainder

The answer is $2x + 2 + \dfrac{1}{x+1}$

Chapter 10: Factoring Polynomials

Greatest Common Factor:

Factoring means breaking something into smaller parts. The easiest way to factor polynomials is by finding the **greatest common factor** (GCF). The GCF is the largest value (constant, variable, or both) that is common to every term in the polynomial.

Example

Factor the following polynomial:
$$2x^5 + 8x^3 + 16x^2 + 4x$$

Every term is evenly divisible by 2, so that is one factor.

Every term also has at least one x, making that another factor.

The GCF is then $2x$. The polynomial can be written:

$$2x(x^4 + 4x^2 + 8x + 2)$$

Example

Factor the following polynomial:
$$9x^4 + 6x^3 + 3x^2$$

Every term is evenly divisible by 3, so that is one factor.

Every term is also divisible by x^2, making that another factor.

The GCF is then $3x^2$. The polynomial can be written:
$$3x^2(3x^2 + 2x + 1)$$

Factoring Into Binomials:

Factoring a polynomial into binomials is a little harder, but once you recognize the patterns, it can become second nature.

In the most general terms, our goal is to find two binomials that multiplied together give us a trinomial:

$$ax^2 \pm bx \pm c = (\ x \pm\ )(\ x \pm\ )$$

Here, we are using $a, b,$ and c to represent positive real numbers. These numbers will be given to us. The x is still our variable.

Let's first talk about signs:
- If the sign in front of the c is a plus, then both binomials will have the same operator. That operator will be the one in front of b.
- If the sign in front of c is negative, then one binomial will have a plus, the other will have a minus.

$$ax^2 + bx + c = (\ x +\ )(\ x +\ )$$
$$ax^2 - bx + c = (\ x -\ )(\ x -\ )$$
$$ax^2 \pm bx - c = (\ x +\ )(\ x -\ )$$

Now we need to fill in the blanks with numbers.
Remember our FOIL method to multiply two binomials:

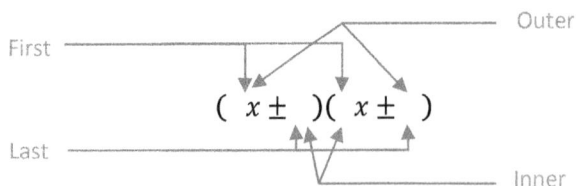

First — — — — — — — Outer

$$(\ x \pm\ )(\ x \pm\ )$$

Last — — — — — — — Inner

FIRST:
$$(\_x)(\_x) = ax^2$$

OUTER + INNER:
$$(\_x)(\_) + (\_)(\_x) = bx$$

LAST:
$$(\_)(\_) = c$$

From practicing the technique, we know:
- a comes from **FIRST**
- b comes from **OUTER + INNER**
- c comes from **LAST**

So, we need to factor a and c, then find some combination of those factors to get b. There's one more rule to make things easier:

$$ax^2 \pm bx + c \rightarrow$$
Factors for a and c ADD to get $|b|$

$$ax^2 \pm bx - c \rightarrow$$
Factors for a and c SUBTRACT to get $|b|$,
The larger product gets the operator in front of b.

Knowing all of this we can now find our binomials. Let's factor:

$$x^2 + 5x + 4$$

Where $a = 1, b = 5$, and $c = 4$. The sign in front of the 4 is a plus, so both binomials will have the same sign. The sign in front of the 5 is also a plus, so both binomials will have a plus sign:

$$(\ x+\ )(\ x+\ )$$

Our next step is to factor a and c. I like to write their factors on either side of the equation:

$$1,1 \diagdown\, x^2 + 5x + 4 \diagup\, \begin{matrix} 4,1 \\ 2,2 \end{matrix}$$

We are calling out that $a = 1 \times 1$, and $c = 4 \times 1$ or 2×2. Because a only has one set of factors, we can write in the FIRST terms:

$$(\, 1x+\ )(\, 1x+\ )$$

To get c, the LAST terms could be 4 and 1 OR 2 and 2. We know because of the last rule that the factors have to add to get $b = 5$ (OUTER + INNER terms): $1 \times? + 1 \times? = 5$

The only pair that works is 4 and 1. Our answer is:

$$(x + 4)(x + 1)$$

To check: $x^2 + x + 4x + 4 = x^2 + 5x + 4$

Example

Factor the following:
$$x^2 - 5x + 6$$

Based on the signs, we want binomials of the form:
$$(\ x - \ )(\ x - \ )$$

Writing out the factors for a and c:

$$1,1 \diagdown x^2 - 5x + 6 \diagdown \begin{matrix} 6,1 \\ 3,2 \end{matrix}$$

We ask what combination of factors ADD (there is a plus in front of c) to get us $|b| = 5$. The answer is $(1 \times 3) + (1 \times 2)$. This give us our factors:

$$(x - 3)(x - 2)$$

Check: $x^2 - 2x - 3x + 6 = x^2 - 5x + 6$

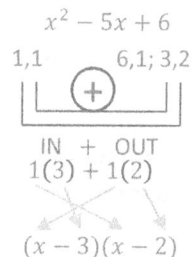

$x^2 - 5x + 6$

1,1   6,1; 3,2

$\boxed{+}$

IN + OUT
$1(3) + 1(2)$

$(x - 3)(x - 2)$

Showing where the factors go.

Example

Factor the following:
$$2,1 \diagup 2x^2 - 4x - 6 \diagdown \begin{matrix} 6,1 \\ 3,2 \end{matrix}$$

Based on the signs, we want binomials of the form:
$$(\ x - \ )(\ x + \ )$$

We already know the factors for a and c, so we ask ourselves what combination of factors SUBTRACT to get $|b| = 4$. The answer is $(1 \times 6) - (2 \times 1)$.

The larger product (the one with the 6) gets the minus sign. This give us our factors:

$$(2x - 6)(x + 1)$$

Check: $2x^2 + 2x - 6x - 6 = x^2 - 4x - 6$

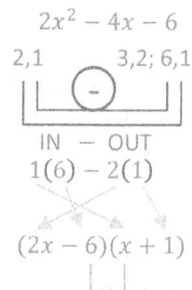

$2x^2 - 4x - 6$

2,1   3,2; 6,1

$\boxed{-}$

IN − OUT
$1(6) - 2(1)$

$(2x - 6)(x + 1)$

Larger product gets the minus sign

Factoring With Equations:

Whenever you see an equation with a power of x higher than 1, you will likely end up with more than one solution. Factoring helps find all of the solutions to an equation directly. To solve:

- **Get all terms on the left side**
- **Factor the left side**
- **Set each factor equal to 0**

Example

Solve the equation for x:
$$x^3 = 5x^2 - 4x$$

Step 1: Get all terms on the left side:
$$x^3 - 5x^2 + 4x = 0$$

Step 2: Factor the left side:
$$x^3 - 5x^2 + 4x = 0$$

$$x(x^2 - 5x + 4) = 0 \qquad \text{GCF} = x$$

$$x(x - 4)(x - 1) = 0 \qquad \text{Factor into Binomials}$$

$x^2 - 5x + 4$

1,1   2,2; 4,1

IN + OUT
$1(4) + 1(1)$

$(x - 4)(x - 1)$

Showing where the factors go.

Step 3: Set each factor equal to 0:
The factored equation above is true if any of the terms is equal to 0. So x could be:

$$x = 0$$
$$x - 4 = 0 \text{ so } x = 4$$
$$x - 1 = 0 \text{ so } x = 1$$

The solutions are: $x = \{0, 1, 4\}$

Note that this example is similar to the last example of Chapter 6. We have broken the problem into parts to find all of the solutions.

The Quadratic Equation:

There are times when factoring a trinomial doesn't produce a "nice" or "neat" solution. In such cases, you can use the **quadratic equation**:

A 2nd degree polynomial such as $ax^2 + bx + c$ is called a quadratic expression.

If $ax^2 + bx + c = 0$,

Then $x = \dfrac{-b \pm \sqrt{b^2 - 4ac}}{2a}$

Example

Solve the equation for x:
$$2x^2 = 6x - 2$$

Set everything equal to 0: $2x^2 - 6x + 2 = 0$

Use the quadratic equation:
$$x = \frac{-b \pm \sqrt{b^2 - 4ac}}{2a} = \frac{+6 \pm \sqrt{36 - 4(2)(2)}}{2(2)}$$

$$= \frac{6}{4} \pm \frac{\sqrt{36 - 16}}{4} = \frac{3}{2} \pm \frac{\sqrt{20}}{4} = \frac{3}{2} \pm \frac{\sqrt{4(5)}}{4} = \frac{3}{2} \pm \frac{2\sqrt{5}}{4}$$

$$= \frac{3}{2} \pm \frac{\sqrt{5}}{2} = \frac{1}{2}\left(3 \pm \sqrt{5}\right); \quad x = \{2.62, 0.38\}$$

Example

Solve the equation for x:
$$x^2 = 2 - 4x$$

Set everything equal to 0: $x^2 + 4x - 2 = 0$

Use the quadratic equation:
$$x = \frac{-b \pm \sqrt{b^2 - 4ac}}{2a} = \frac{-4 \pm \sqrt{16 - 4(1)(-2)}}{2(1)}$$

$$= -2 \pm \frac{\sqrt{16 + 8}}{2} = -2 \pm \frac{\sqrt{24}}{2} = -2 \pm \sqrt{6}$$

$$x = \{0.45, -4.45\}$$

Chapter 11: Graphing

Equations with Two Variables:

Equations and algebraic expressions are not limited to just one variable. Many times, we need to have two or more variables to solve a problem. Consider the example:

$$4x + 2y = 8$$

There are many solutions to this problem. To illustrate this, we isolate terms. Conventionally, people try to find an equation that looks like $y = \cdots$. You don't have to do it this way, but since everyone else does, that's what we'll do:

$$4x + 2y = 8$$
$$2y = 8 - 4x$$
$$y = 4 - 2x$$

Now we can see more clearly that there are many solutions. If $x = 0, y = 4$. If $x = 1, y = 2$. In fact, we can make a table of solutions like this:

x	-3	-2	-1	0	1	2	3
y	10	8	6	4	2	0	-2

We can let x take on any value we want and then find what the corresponding y value is. Here we let x be any integer from -3 to 3. Each column in the table constitutes an **ordered pair** - a pair of numbers listed in order of x value then y value. These ordered pairs are written:

$$(x, y) = (-3, 10) \qquad (x, y) = (1, 2)$$
$$(x, y) = (-2, 8) \qquad (x, y) = (2, 0)$$
$$(x, y) = (-1, 6) \qquad (x, y) = (3, -2)$$
$$(x, y) = (0, 4)$$

Graphing:

The reason we write the solutions of a two variable system as ordered pairs is because we want to **graph** the solution. To graph a solution, we do the following:

1. Draw a horizontal number line that represents the x values. This is called the **x-axis**.
2. Draw a vertical number line that represents the y values. This is called the **y-axis**.
3. Now **plot** the ordered pairs, i.e. put a dot at the x and y value of each ordered pair.
4. Draw a smooth line connecting the dots.

For our example, the graph looks like this:

$(x, y) = (-3, 10)$

$(x, y) = (-2, 8)$

$(x, y) = (-1, 6)$

$(x, y) = (0, 4)$

$(x, y) = (1, 2)$

$(x, y) = (2, 0)$

$(x, y) = (3, -2)$

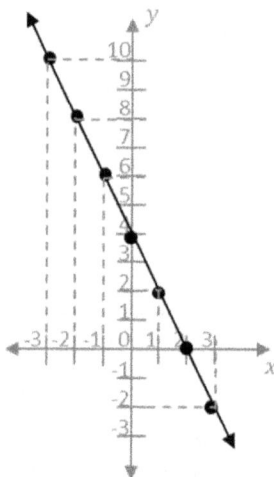

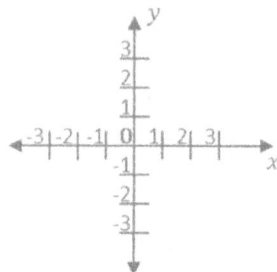

The x and y number lines for a graph. The axes are labeled and the double arrows indicate that each number line goes from $-\infty$ to ∞.

Notice a few of things:
*   Both number lines have double arrows indicating that x and y can have values from $-\infty$ to ∞.
*   The smooth line connecting the dots has double arrows showing that the solution set goes from $-\infty$ to ∞.
*   The smooth line itself shows that ANY set of ordered pairs ON the line is a valid solution. For example, if $x = 1.5$, then $y = 1$. We only plotted integer values of x because it's easier than plotting an infinite number of points.

There are a few more terms associated with graphing that you should know. The **independent variable** is the one we get to pick values for at random. In our example, x is the independent variable.

The **dependent variable** takes on values dictated by the equation and the values for the independent variable. The dependent variable, y in our case, is the one isolated to the left of the equation ($y = \cdots$).

The span of x values is called the **domain**, and the resulting span of y values is the **range**. We plot the x and y values in the $x - y$ **plane**. A plane is a flat surface that extends from $-\infty$ to ∞ in every direction.

Graphing is nice because we can read the solutions from the picture rather than calculating them from the equation. For complicated equations, this can be a real time saver.

So how many points do you have to plot to get a good graph? It depends on the equation. There are several types of well known equations with distinctive forms that you will learn about in Algebra 2. For now, we will usually deal with lines, and lines only need two points for a good graph. If you have to plot an equation with powers of x and/or y greater than 1, plot a bunch of points.

Example

Graph: $4x - y = 2$

First we get our correct form:
$$4x - y = 2$$
$$-y = 2 - 4x$$
$$y = 4x - 2$$

Let's pick $x = 0$, and $x = 1$.
   When $x = 0, y = 4(0) - 2 = -2$.
   When $x = 1, y = 4(1) - 2 = 2$.

The graph is to the right.

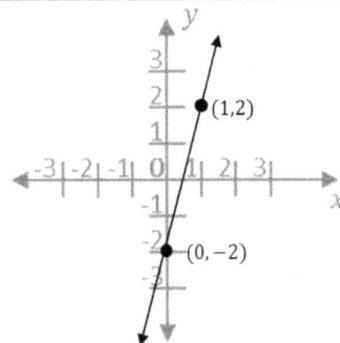

Graph of
$4x - y = 2$

3-D Graphs:

Graphs of equations with two variables are called **two dimensional** (2-D) graphs. Graphs of equations with three variables are called **three dimensional** (3-D) graphs. Three variable equations have two independent variables that make up the domain and one dependent variable that makes up the range.

To start a 3-D plot, we need to set up the **Rectangular** or **Cartesian Coordinate system**. We plot three number lines, one for each variable, so that they are squared against each other. In doing this, we must obey the **Right Hand Rule**:

> **Point the thumb of your right hand in the +z direction, your straight fingers point in the +x direction, and bending your fingers will point in the +y direction.**

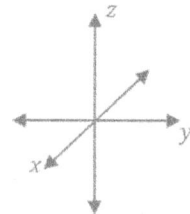

The Right Handed Cartesian Coordinate System. Labels are on the positive side of the axes. Positive x values come out of the page.

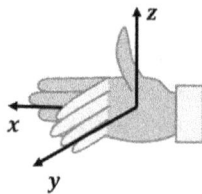

The Right Hand Rule.

We now have three planes: the $x - y$ plane, the $x - z$ plane, and the $y - z$ plane as shown.

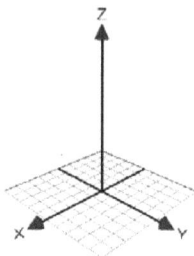

$x - y$ plane

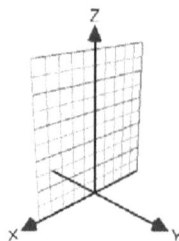

$x - z$ plane

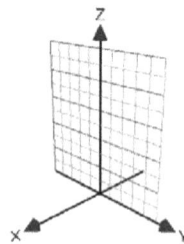

$y - z$ plane

Example

Graph the following equation:
$$z = x^2 + y^2$$

Now we have to find **ordered triplets**. We pick x and y values (the domain) at random to come up with z values (the range):

x	-1	-1	-1	0	0	0	1	1	1
y	-1	0	1	-1	0	1	-1	0	1
z	2	1	2	1	0	1	2	1	2

Examples of points on this graph are:
$(x, y, z) = (-1, -1, 2), (-1, 0, 1), (0, -1, 1)$, etc.

And, the 3-D graph is shown to the right.

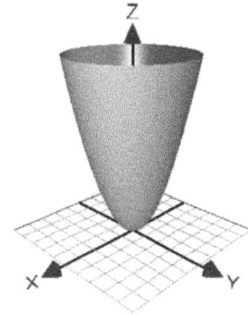

Graph for
$z = x^2 + y^2$

Overview of Contour Plots:

Sometimes graphing in 3-D is a pain, or we have equations with more than three variables, In these cases, we can hold one or more variables constant and then plot the resulting equations in 2- or 3-D. These graphs are called **contour plots**.

In the above example, we had an equation with three variables which forced us to do a 3-D graph. If we set z equal to some constant, we reduce the equation to a two variable problem. For example, if we assign $z = 1$ in the equation $z = x^2 + y^2$, we get the equation:

$$x^2 + y^2 = 1$$

All ordered pairs that solve this equation should form a 2-D graph of what is happening in 3-D at $z = 1$. In other words, we should be able to graph a slice, or **cross-section**, of the 3-D graph at a height of $z = 1$. Plotting multiple slices on the same graph is what gives us the contour plot.

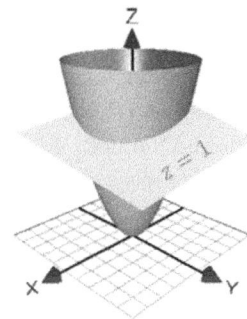

Slicing the graph for
$z = x^2 + y^2$
at the height $z = 1$.

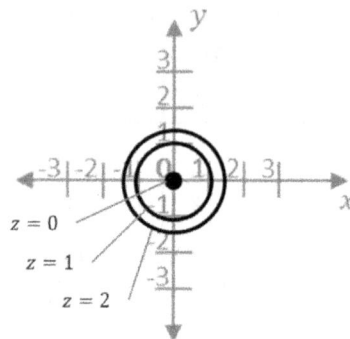

> **Example**
>
> Draw the contour plot of:
> $$z = x^2 + y^2$$
>
> From the example on the prior page, we can read from the table:
> - When $z = 2$, $x^2 + y^2 = 2$, and we get $(x, y) = (-1, -1), (-1, 1), (1, -1)$, and $(1, 1)$.
>
> - When $z = 1$, $x^2 + y^2 = 1$, and we get $(x, y) = (-1, 0), (0, -1), (1, 0)$, and $(0, 1)$.
>
> - When $z = 0$, $(x, y) = 0$.
>
> If we graph the solutions of each of the above equations on the same $x - y$ plane, we get the contour plot to the right.
>
> Comparing this to the 3-D graph of the prior page, we can see that we are plotting the slices, or cross-sections of the 3-D graph at different z values.

Contour Plot for
$z = x^2 + y^2$

Note: Though most Algebra 1 courses don't cover contour plots, they are included here briefly for completion.

Also, in Algebra 2, you will learn to recognize equations of circles like those in the above example. For now, when presented with equations with powers of x and/or y higher than 1, assume a curved line and plot a bunch of points.

Chapter 12: Lines

Distance, Midpoint, & Slope:

Let's look at lines in a bit more detail. As we noted in the previous chapter, lines can be defined by at least two points. The two points can be written as (x_1, y_1) and (x_2, y_2). The y_1 and y_2 are the values of y we get from an equation after plugging in x_1 and x_2 values for x.

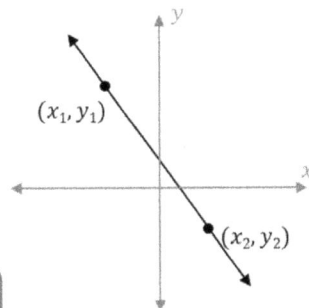

A line defined by two points: (x_1, y_1) and (x_2, y_2).

The **distance** between the two points is given by:

$$distance = d = \sqrt{(x_2 - x_1)^2 + (y_2 - y_1)^2}$$

The **midpoint** between the two points is given by:

$$midpoint = \left(\frac{x_1+x_2}{2}, \frac{y_1+y_2}{2}\right)$$

Example

Find the distance and midpoint between:
$$(1, -2) \text{ and } (3,4)$$

Plugging these points into the formulas we get:

$$d = \sqrt{(x_2 - x_1)^2 + (y_2 - y_1)^2}$$
$$= \sqrt{(3 - 4)^2 + (-2 - 1)^2}$$
$$= \sqrt{(-1)^2 + (-3)^2} = \sqrt{1 + 9} = \sqrt{10}$$

$$midpoint = \left(\frac{x_1+x_2}{2}, \frac{y_1+y_2}{2}\right)$$
$$= \left(\frac{1+3}{2}, \frac{-2+4}{2}\right) = \left(\frac{4}{2}, \frac{2}{2}\right) = (2,1)$$

Don't worry about choosing which is point 1 and which is point 2. As long as you are consistent (don't mix up the x's or y's), it doesn't matter.

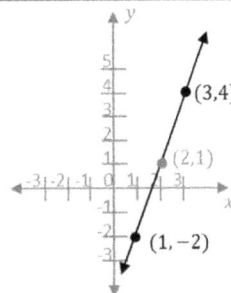

Given points are in black, midpoint is in grey.

If we switch the labeling of the points, we get the same result:

$$d = \sqrt{(4 - 3)^2 + (1 - (-2))^2}$$
$$= \sqrt{(1)^2 + (3)^2} = \sqrt{10}$$

$$midpoint = \left(\frac{3+1}{2}, \frac{4+(-2)}{2}\right)$$
$$= \left(\frac{4}{2}, \frac{2}{2}\right) = (2,1)$$

When we talk about lines, we often want to know the **slope** of the line, i.e. the direction the line is heading if traveling from left to right. Slope is more formally described by the change in height (Δy) over the change in length (Δx). The slope between two points is given by:

$$slope = m = \frac{\Delta y}{\Delta x} = \frac{y_2 - y_1}{x_2 - x_1}$$

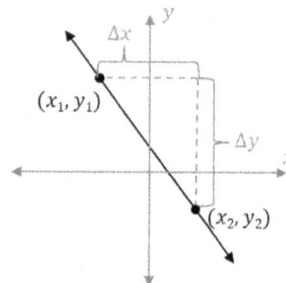

Calculating Slope

It does not matter which point you pick as (x_1, y_1) or (x_2, y_2) as long as you are consistent (don't mix the x's and y's between pairs). Also, you can pick any two points on a straight line to calculate the slope.

Example

Find the slope between:

$$(1, -2) \text{ and } (3, 4)$$

Plugging these points into the formula we get:

$$m = \frac{\Delta y}{\Delta x} = \frac{y_2 - y_1}{x_2 - x_1} = \frac{-2 - 4}{1 - 3} = \frac{-6}{-2} = 3$$

If we switch the order of the points, we get the same result:

$$m = \frac{4 - (-2)}{3 - 1} = \frac{6}{2} = 3$$

A slope of +3 means that the line points upward (positive) and has a rise to run ratio of 3 (goes up a distance of 3 for every 1 in length).

Equations for Lines:

What is nice about slope is that it allows us to get the equation for a line. Let's imagine that we know the slope of the line, m, but we only know one point (a, b). Let's plug these into the slope equation:

$$m = \frac{y_2 - y_1}{x_2 - x_1} = \frac{y - b}{x - a}$$

Notice that we dropped the subscripts for the point (x_2, y_2) above since we no longer need them.

If we now solve for $y - b$, we get:

$$y - b = m(x - a)$$

This is the **Point-Slope Form** for the equation of a line because it contains a point and a slope.

Find the equation for the line between:
$$(1, -2) \text{ and } (3,4)$$

We already know the slope is 3 from the prior example. Pick one of the points above for (a, b), and plug it into the point-slope form:

$$y + 2 = 3(x - 1)$$

If we were to plug in $(x, y) = (3,4)$, the equation should be true:
$$4 + 2 = 3(3 - 1)$$
$$6 = 6$$

The point-slope form of an equation of a line can give us any point on the line.

Equivalently, we could have chosen $(a, b) = (3,4)$:
$$y - 4 = 3(x - 3)$$

If we plug in $(x, y) = (1, -2)$, we get:
$$-2 - 4 = 3(1 - 3)$$
$$-6 = -6$$
which is true.

Now consider the special case of choosing (a, b) as the point where the line crosses the y-axis. This forces $a = 0$, so the point is now $(0, b)$. In this case, b has a special name and is called the **y-intercept** of the line.

Plugging this point into the point-slope form of the equation for a line gives:
$$y - b = m(x - 0)$$

Solving for y gives:

$$y = mx + b$$

This is the **Slope-Intercept Form** for the equation of a line as it contains the slope and the y-intercept.

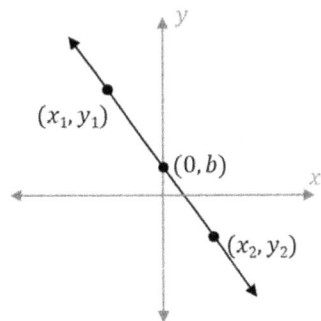

The y-intercept of a line is the point where the line crosses the y-axis, i.e. where $x = 0$.

> **Example**

Find the slope-intercept equation for the line between:
$$(1, -2) \text{ and } (3,4)$$

We already have one equation for the line:
$$y + 2 = 3(x - 1)$$

If we solve this equation for y, we get:
$$y + 2 = 3x - 3$$
$$y = 3x - 5$$

So, the y-intercept occurs at (0,-5).

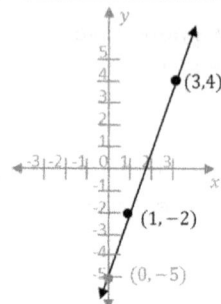

The y-intercept is in grey.

Note: The equation
$$y - 4 = 3(x - 3)$$
would give the same result.

> **Example**

Plot the line given by:
$$2y - 4x = -4$$

First we put the equation in slope-intercept form:
$$2y = 4x - 4$$
$$y = 2x - 2$$

This equation tells us that the slope is 2, and the y-intercept is -2. We plot the point $(0, -2)$, then go up two and over one $(m = +2)$ for a second point. The line connects the dots.

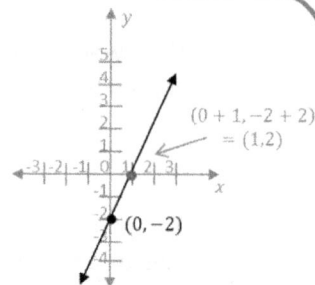

The y-intercept is in black, the second point determined from the slope is in grey.

A final note on lines:

For a **horizontal line** (back and forth), all y's are equal, so $\Delta y = y_2 - y_1 = 0$ which forces $m = {}^{\Delta y}/_{\Delta x} = 0$. The equation for the line becomes $y = b$.

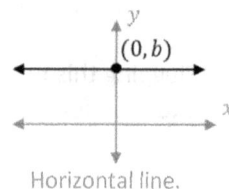

Horizontal line.

For a **vertical line** (up and down), all x's are equal, so $\Delta x = x_2 - x_1 = 0$ which forces $m = {}^{\Delta y}/_{\Delta x} = \infty$. There is no point-slope or slope-intercept form for the equation of a vertical line. The equation is simply $x = c$.

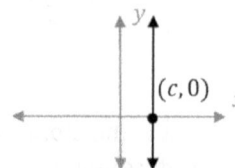

Vertical line.

Chapter 13: Systems of Equations

Systems of Independent Equations:

An interesting problem arises when we are given two or more equations to solve at once, i.e. a **system of equations**. We can come up with solutions for each equation, but what we really want to know is what solutions are common to all of the equations. This is referred to as **simultaneously solving** the system of equations.

For a two equation system, if S_1 is the solution set for one equation, and S_2 is the solution set for the second equation, then we want to find:

$$S_1 \cap S_2$$

There are two things we need to check before even attempting to do this:
1. We have as many variables as we have equations (people often say, "as many equations as unknowns").
2. All of the equations are **independent** of each other.

The first criterion is easy enough to check; if you have two equations, you better have two variables. The second criterion calls for the equations to be independent. This means that you can't derive one equation from another.

When you solve single equations, you perform operations to both sides of the equals sign. If you can do this to one equation to get the other equation, then the two equations are **dependent** on one another. For example:

$$y = 2x + 4 \text{ and } 2y = 4x + 8$$

are dependent equations because you can multiply the first equation by 2 to get the second equation.

However, if you have successfully met the criteria above, then you can proceed to simultaneously solve the system using any of three following techniques:

Combination, Substitution, or Graphing

Combination:

The first method to simultaneously solve a system of equations is to combine them through addition or subtraction to get rid of one variable. This gives a third equation with one unknown which can then be solved. Plugging that solution into either of the original two equations will give the values for the other variable.

Example

Solve the system of equations:
$$y = 4x + 3; \quad 2y = 4x - 2$$

Let's try to get rid of the y variable by subtracting 2 times the first equation from the second equation:

$$\begin{array}{r} 2y = 4x - 2 \\ - \ 2 \times [y = 4x + 3] \end{array} \quad \Longrightarrow \quad \begin{array}{r} 2y = 4x - 2 \\ + \quad - 2y = -8x - 6 \\ \hline 0 = -4x - 8 \end{array}$$

Solving $0 = -4x - 8$ gives us $x = -2$. Plugging that result back into the first equation gives:
$$y = 4(-2) + 3 = -5.$$

Or, you can also plug $x = -2$ into the 2nd equation:

$$2y = 4(-2) - 2 = -10$$
$$y = {}^{-10}/_2 = -5$$

The solution to both equations is $(x, y) = (-2, -5)$.

Example

Solve the system of equations:
$$3y = x + 2; \quad y = x - 1$$

Get rid of the x variable by subtracting the first equation from the second:

$$\begin{array}{r} 3y = x + 2 \\ - \ [y = x - 1] \\ \hline 2y = \quad 3 \end{array}$$

Solving $2y = 3$ gives us $y = \frac{3}{2}$. Plugging this result into the second equation gives:
$$\frac{3}{2} = x - 1 \ \rightarrow \ x = 1 + \frac{3}{2} = \frac{5}{2}$$

The solution to both equations is $(x, y) = (\frac{5}{2}, \frac{3}{2})$.

Substitution:

The second method to simultaneously solve a system of equations is to use substitution:
- Isolate one variable in the first equation so that you get an expression for that variable ($y = \cdots$).
- Substitute the expression found in step one into all occurrences of the variable in the second equation.
- You now have one equation with one unknown. Solve this equation to find the values for this variable.
- Plug the result back into the first equation to find the values for the first variable.

Example

Solve the system of equations:
$$y = 4x + 3; \quad 2y = 4x - 2$$

The first equation is already solved in terms of one variable:
$$y = 4x + 3$$

So, we can plug in $(4x + 3)$ for y in the second equation and solve:
$$2y = 4x - 2$$
$$2(4x + 3) = 4x - 2$$
$$8x + 6 = 4x - 2$$
$$x = -2$$

Plug this into either of the first equations and get:
$$y = 4(-2) + 3 = -5$$

Example

Solve the system of equations:
$$3y = x + 2; \quad y = x - 1$$

Plug in the second equation into the first:
$$3(x - 1) = x + 2$$
$$2x = 5 \rightarrow x = \frac{5}{2}$$

Plugging back into the second equation:
$$y = \frac{5}{2} - 1 = \frac{3}{2}$$

Graphing Solutions:

The third method to simultaneously solve a system of equations is to graph both equations and literally look for the intersections. The drawback is that the solution is only as accurate as your drawing.

Example

Solve the system of equations:
$$y = 4x + 3; \quad 2y = 4x - 2$$

The first equation is in slope-intercept form:
$$y = 4x + 3$$

We plot the point $(0,3)$, then use the slope, $m = 4$, to plot the line.

The second equation can be modified to fit into slope-intercept form:
$$2y = 4x - 2$$
$$y = 2x - 1$$

We plot the point $(0, -1)$, then use the slope, $m = 2$, to plot the line.

The plot for both equations is to the right. By inspection, we can see that they overlap at the point $(x, y) = (-2, -5)$.

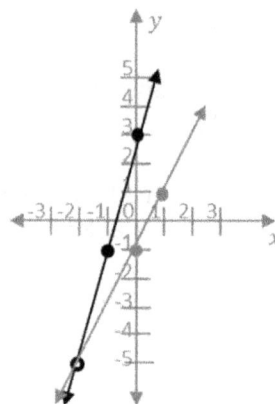

The first line is in black, the second in grey. The intersection is circled.

Example

Solve the system of equations:
$$3y = x + 2; \quad y = x - 1$$

Putting both equations into slope-intercept form:
$$y = \frac{x}{3} + \frac{2}{3};$$

$$y = x - 1$$

Graphing the equations gives the intercept at:

$$(x, y) = \left(\frac{5}{2}, \frac{3}{2}\right)$$

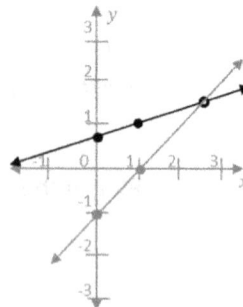

The first line is in black, the second in grey. The intersection is circled.

Chapter 14: Word Problems

Cracking the Code:

Many students are afraid of word problems, but in truth you have been solving such problems all of your life. If someone came to you and said, "I have 10 candies, do you want half of them?," you're going to let them know it if you get less than five.

The key to success with word problems, just as it is in any math problem, is recognizing patterns. Certain key words imply certain operations. Once you figure out the code, you're half way there. So, why figure it out if I can write it down for you? Here it is:

Operator/ Expression	Key Words
x	What, How many, How much, Any *Noun* can be a variable
$=$	Is, Are, Does, Do, Was, Were, Yields, Same As, Cost
$+$	And, Increased, More Than, Combined, Total, Sum
$-$	Decreased by, Less Than, Fewer Than, Difference, From, Left
$\times$	Of, Product, Factor of
$\div$	Into, Out of, Per, Any Statement of Ratio, Fraction, or Percentage

Once you know the key words, it's easy to set up the problem. Once you set up the problem, you solve it like any other problem.

Examples:

Percentages are easy when thinking of them as a word problem. What is 40% of 58?:

$$x = \frac{40}{100} \times 58 = 23.2$$

My car just went 50 miles on three gallons of gas. How many miles per gallon is that?

The last question is:
How many miles per gallon is that?

$$x = miles \div gallon$$

$$x = \frac{50}{3} = 16.7$$

There was half a pie in the refrigerator, and now two-thirds of that is gone. How much is left?

Rephrasing: What is left from ½ a pie after 2/3 of it is gone?

$$x = \frac{1}{2} - \left(\frac{1}{2} \times \frac{2}{3}\right)$$

$$= \frac{1}{2} - \frac{2}{6} = \frac{1}{6}$$

Is it true that three will go evenly into the sum of any three consecutive integers?

Call the first integer x. The next integer is $x + 1$. The next one is $x + 2$. "Sum" means add them together:
$$x + x + 1 + x + 2 = 3x + 3.$$

"Into" means divide - Since $3x + 3$ is evenly divisible by 3, the statement is true.

A square has the same area as a rectangle whose one side is half of the square's side and whose other side exceeds the square's side by 3. How big is the side of the square?

Let x be the side of the square. The area of the square is then $x \times x = x^2$. The rectangle sides are:

One side is half of the square's side: $\frac{1}{2} \times x$

One side exceeds x by three: $x + 3$

The area of the rectangle is $\frac{x}{2}(x + 3) = \frac{x^2 + 3x}{2}$.

The problem states both areas are equal, so setting them equal and solving for x gives:

$$x^2 = \frac{x^2 + 3x}{2}$$

$$2x^2 = x^2 + 3x$$
$$x^2 = 3x$$
$$x = 3$$

Note, $x = 0$ is also a solution to the *equation*, but it doesn't make sense to the *word problem*.

Three cans of cola and four bags of chips cost $30.50. Two cans of cola and eight bags of chips costs $51.00. How much do cans of cola and bags of chips cost each?

Let the cost of cola be x, the cost of chips be y. We have two unknowns, so we need two equations:

$$3x + 4y = 30.5; \quad 2x + 8y = 51$$

Solving the second equation for x gives:
$$2x = 51 - 8y \quad \rightarrow \quad x = 25.5 - 4y$$

Plugging into the first equation gives:
$$3(25.5 - 4y) + 4y = 30.5$$
$$-8y = -46$$
$$y = \$5.75 \text{ for chips}$$

Plugging into $x = 25.5 - 4y$ gives $x = \$2.50$ for soda

Appendices

A: Course Summary
B: Problem Sets
C: Solutions to Problem Sets

Appendix A: Course Summary

1   **Review** of key concepts:
      There are many types of **numbers**:
         Real, positive, negative, integers, rational, irrational, prime, compound
         Reviewed how to work with negative numbers
         Factors break compound numbers into prime subcomponents

2   **Fractions** can be manipulated by multiplying by 1
         Reviewed Reduction, Rationalization, Addition, Multiplication
         Reviewed Proportion, Ratio, and Percentage

3   **Exponents** make bases operate on themselves
         Positive powers multiply
         Negative powers invert
         Fractional powers multiply and take roots
         Like bases multiply through addition of exponents
         Like bases divide through subtraction of exponents

      Order of Operations & **Properties**:
         Parentheses, Exponents, Multiplication, Addition
         Commutative, Associative, Distributive, Identity, Inverse

4   **Solving Equations**:
      What you do to one side, you do to the other
           OR Add or Multiply by true expressions

5   Typically reverse order of operations to simplify equation solving
      Absolute value sign creates two problems to solve

6   Isolate variable
      Group like terms

7   **Sets:**
      Notation
      Complement, Union, Intersection
      Useful way to express more than one solution

8   Solving **Inequalities**:
      Addition (subtraction) the same
      Multiplication by negative number flips inequality
      Absolute value sign creates two problems to solve

9 **Polynomials**:
Addition (subtraction) – Group like terms
Multiplication – Multiply all terms together like multi-digit multiplication
  FOIL- First Inner Outer Last
Divide – Use Fractions or long division

10 **Factoring Polynomials**:
GCF- **Greatest Common Factor**
Reversing FOIL to factor trinomials $ax^2 \pm bx \pm c$:

$ax^2 \pm bx + c$ $ax^2 \pm bx - c$

Add to get $|b|$ Subtract to get $|b|$

Binomials both get   Different signs, Largest
sign in front of b product gets sign in front of b

In equations – get all terms to left, set each factor equal to 0
Quadratic Equation:
$$x = \frac{-b \pm \sqrt{b^2 - 4ac}}{2a}$$

11 **Graphing**:
Create ordered pairs (points) by isolating the dependent variable and solving for random values of independent variables.
Contour plots set one or more variables equal to a constants.

12 **Lines have two equations**:
  Point-Slope: $y - b = m(x - a)$
  Slope-Intercept: $y = mx + b$

13 **Systems of Equations**:
Make sure # equations = # unknowns
Make sure equations are independent of each other

Solve by Combination, Substitution, or Graphing

14 **Word Problems**:
Recognize key words

Chapter correlations are given in large grey font.

Appendix B: Problem Sets

Chapter 1

Find all of the factors for the following:

1.1: 249

1.2: 512

1.3: 1017

1.4: 52

1.5: 100

1.6: 88

1.7: 99

1.8: 45

Evaluate:

1.9: $5 + (-9) - (-4)$

1.10: $2 + 8 - (-3)$

1.11: -2×8

1.12: $-3 \times -4 \times -2$

1.13: $-14/(-7)$

1.14: $-14/|-7|$

1.15: $-|22|/|-2|$

1.16: $-10 - |-5|$

Chapter 2

Reduce or Rationalize the following:

2.1: $\dfrac{52}{4}$ **2.2:** $\dfrac{12}{56}$

2.3: $\dfrac{2}{\sqrt{2}}$ **2.4:** $\dfrac{100}{\sqrt{20}}$

2.5: $\dfrac{300}{3\sqrt{10}}$ **2.6:** $\dfrac{24}{2\sqrt{8}}$

Evaluate:

2.7: $\dfrac{2}{3} + \dfrac{4}{5}$ **2.8:** $\dfrac{3}{2} + \dfrac{-5}{6}$

2.9: $\dfrac{5}{12} - \dfrac{4}{20}$ **2.10:** $\dfrac{2}{3} - \dfrac{3}{4} + \dfrac{1}{6}$

2.11: $-\dfrac{3}{2} \times \dfrac{2}{5}$ **2.12:** $\dfrac{3}{4} \times \dfrac{1}{12} \times \dfrac{2}{3}$

2.13: $\dfrac{1}{4} \div \dfrac{7}{8}$ **2.14:** $\dfrac{2}{6} \div \dfrac{-1}{7}$

Convert:

2.15: 2500 in^3 to ft^3 **2.16:** $20\dfrac{\text{feet}}{\text{second}}$ to $\dfrac{\text{miles}}{\text{hour}}$

Chapter 3

Evaluate:

3.1: $5^2 \times 5$

3.2: $2^2 \times 2^4$

3.3: $4^{-2} \times 4^2$

3.4: $4^2 \times 2^3$

3.5: $8^3 \div 8^4$

3.6: $4^2 \div \sqrt{4}$

3.7: $\frac{1}{7} \times 7^3$

3.8: $4^{3/2} \times \frac{1}{2}$

3.9: $5 \times (2 - 3)$

3.10: $3 \times \left(4 - (8 + 2)\right)$

3.11: $(2^2 + 3)^2$

3.12: $\left(\frac{18}{(4-2)}\right)\left(\frac{(-2)^2}{6}\right)$

3.13: $|(-2)^3(4 - 3)|$

3.14: $\sqrt{(4 \times 2^4)}$

3.15: $\sqrt[3]{5} \times |-12 + 7|^{-2}$

3.16: $((4 - 2)^{(1+2)} \times 8^2)^{-1/2}$

Chapter 4

Solve for x:

4.1: $x + 5 = 2$ **4.2:** $x - 3 = 4$

4.3: $x + 9 = 10$ **4.4:** $x - 2 = 8$

4.5: $x + 2 = 13$ **4.6:** $x - 3 = 9$

4.7: $4x = 3$ **4.8:** $2x = 7$

4.9: $5x + 2 = -3$ **4.10:** $4x + 5 = 9$

4.11: $2x + 9 = 15$ **4.12:** $3x - 2 = 7$

4.13: $4x - 1 = 3$ **4.14:** $2x + 3 = 7$

4.15: $\frac{x}{5} + 3 = 5$ **4.16:** $\frac{2x}{3} - 1 = 3$

Chapter 5

Solve for x:

5.1: $4x^2 - 2 = 14$

5.2: $3x^2 - 3 = 0$

5.3: $\sqrt{x} - 4 = 0$

5.4: $3x^{-2} + \frac{1}{3} = \frac{2}{3}$

5.5: $x^3 + 4 = 12$

5.6: $5x^2 + 5 = 50$

5.7: $4(x + 3) = 20$

5.8: $2(x - 5) = -4$

5.9: $\frac{x-2}{4} + 3 = 4$

5.10: $3\left(\frac{x-2}{4}\right) - 6 = -6$

5.11: $2(x^3 - 2) + 10 = 22$

5.12: $2(3x^2 - 30) + 6 = 42$

5.13: $4|x - 3| = 16$

5.14: $-2|x + 2| + 4 = 2$

5.15: $2(x^2 - 3) + 5 = 1$

5.16: $8(x^{-1/3} + 2) - 20 = 4$

Chapter 6

Solve for x:

6.1: $2x - 3x^2 = -4x^2 + 3x$

6.2: $x^2 - 3x = 2x - 4x^2$

6.3: $3x + 8 = 4x - 2$

6.4: $5x - 2 = 8x + 1$

6.5: $x = 2(x + 7)$

6.6: $4(x + 2) = 2(x + 7)$

6.7: $4x = 3(x - 4)$

6.8: $4 - x = 7(x + 3)$

6.9: $5(x^{-1} + 2) = \frac{3x+4}{x}$

6.10: $\frac{x-2}{3x} = \frac{4}{x} + 3$

6.11: $5(\sqrt{x} + 2) = 3\sqrt{x} + 14$

6.12: $x^2 + 5 = 9(x^2 - 2)$

6.13: $2x^2 + 9 = 3x^2 + 3$

6.14: $x\left(x - \frac{2}{x}\right) = 7(3 - x^2)$

6.15: $x^3 - 1 = 3x(x^2 - \frac{2}{x})$

6.16: $2\sqrt[3]{x} + 3 = 3x^{1/3} + 1$

Chapter 7

Use set-builder notation for the following:

7.1: A is the set of real numbers greater than 100.

7.2: A is the set of real numbers between 25 and 1000.

7.3: A is the set of integers between 0 and 10.

7.4: A is the set of rational numbers greater than 2.

Find A':

7.5: $A = \{x: x > 6; x \in B\}$
$B = \{x: 0 \leq x \leq 10; x \in \mathbb{Z}\}$

7.6: $A = \{x: 3 < x < 7; x \in B\}$
$B = \{x: 0 \leq x \leq 10; x \in \mathbb{Z}\}$

7.7: $A = \{even\ integers; x \in B\}$
$B = \{x: 0 \leq x \leq 10; x \in \mathbb{Z}\}$

7.8: $A = \{x: x > 5; x \in B\}$
$B = \{x: 0 \leq x \leq 10; x \in \mathbb{R}\}$

7.9: $A = \{x: x > 52; x \in B\}$
$B = \{x: 0 \leq x \leq 100; x \in \mathbb{R}\}$

7.10: $A = \{x: 25 < x < 50; x \in B\}$
$B = \{x: 0 \leq x \leq 100; x \in \mathbb{R}\}$

Find $A \cup B$, and $A \cap B$:

7.11: $A = \{x: x > 6; x \in \mathbb{Z}\}$
$B = \{x: 0 \leq x \leq 10; x \in \mathbb{Z}\}$

7.12: $A = \{x: 3 < x < 7; x \in \mathbb{Z}\}$
$B = \{x: 0 \leq x \leq 10; x \in \mathbb{Z}\}$

7.13: $A = \{x: x > 6; x \in \mathbb{R}\}$
$B = \{x: 0 \leq x \leq 10; x \in \mathbb{Z}\}$

7.14: $A = \{x: x < 0; x \in \mathbb{R}\}$
$B = \{x: x > 4; x \in \mathbb{R}\}$

7.15: $A = \{x: 0 < x < 25; x \in \mathbb{R}\}$
$B = \{x: 2 \leq x \leq 10; x \in \mathbb{R}\}$

7.16: $A = \{x: -5 \leq x \leq 5; x \in \mathbb{R}\}$
$B = \{x: x > 0; x \in \mathbb{R}\}$

Chapter 8

Solve for x:

8.1: $2x + 5 > 2$ **8.2:** $7x - 3 < 4$

8.3: $-4x + 9 \geq 1$ **8.4:** $-2x - 2 < 8$

8.5: $-x + 2 \leq 13$ **8.6:** $4x - 3 > 9$

8.7: $-4(x + 2) < 8$ **8.8:** $2(x - 8) < 10$

8.9: $4x + 3 < 5x + 2$ **8.10:** $5 - x \geq 9x - 5$

8.11: $\frac{2x}{3} + 1 > x - 5$ **8.12:** $\frac{x}{5} - 2 \leq x - 3$

8.13: $-4x^2 + 5 < 3x^2 - 2$ **8.14:** $-2(x^3 + 4) > 24$

8.15: $|x - 3| \geq 5$ **8.16:** $|3 - 2x| < 7$

Chapter 9

Evaluate the following:

9.1: $(3x^2 + x + 4) - (x + 2)$ **9.2:** $(x^2 + 4x - 3) - (3x^2 - 4)$

9.3: $(2x^2 + x - 2) - (x + 8)$ **9.4:** $(x^2 + 4x + 1) - (3x + 4)$

9.5: $(2x + 1)(x^2 - 2x + 1)$ **9.6:** $(3x - 3)(x^2 + 2x + 5)$

9.7: $(2x^2 - 3x + 9)(x^2 - x + 3)$ **9.8:** $(x^4 + x^3 + 1)(x^2 + x + 1)$

Use FOIL to multiply the following pairs of expressions:

9.9: $(2x + 1)(x - 1)$ **9.10:** $(x + 3)(x - 5)$

9.11: $(-x - 2)(x + 1)$ **9.12:** $(4x - 7)(2x + 3)$

9.13: $(3x + 1)(3x - 1)$ **9.14:** $(-2x - 4)(4x + 2)$

9.15: $\left(\frac{x}{2} + 3\right)\left(\frac{x}{4} - 2\right)$ **9.16:** $\left(\frac{2x}{3} - 1\right)\left(\frac{x}{3} + 1\right)$

Chapter 10

Solve for x:

10.1: $x^2 + 3x = x - 1$

10.2: $x^2 + 3x = -2$

10.3: $x^3 + 5x^2 = x^2 - 3x$

10.4: $2x^2 + 1 = -5x$

10.5: $x^4 + x^2 = 7x^2 - 5$

10.6: $x^2 = 4x - 4$

10.7: $x^3 = 5x^2 - 6x$

10.8: $3x^2 = 11x - 10$

10.9: $x^2 = 2 - x$

10.10: $-x^2 = 2x - 3$

10.11: $x^2 = 5 - 4x$

10.12: $x^2 + \frac{7x}{2} = 2$

10.13: $x^2 = x + 6$

10.14: $x^2 = 1 + 2x$

10.15: $x^2 = 1$

10.16: $3x^2 = 7x + 6$

Chapter 11

Graph the following in 2-D:

11.1: $2x - y = 5$

11.2: $4x + 2y = 12$

11.3: $y - 3x = 9$

11.4: $y - 2x = 10$

11.5: $y + 4x = 13$

11.6: $4 - y = 3x$

11.7: $y - x^2 = 2$

11.8: $y - 2x^2 + 3(x - 2) = 0$

11.9: $x(x - 3) = y - x$

11.10: $4x + 2x^3 = 2y$

11.11: $(x - 2)^2 = y$

11.12: $x^4 - x^2 = y + 2$

Show a contour plot in 2-D for the following:

11.13: $z = 4x + 2y$

11.14: $z = y + x^2 + 2$

11.15: $z = y - x^3$

11.16: $z = y - 2x^2$

Chapter 12

Find the distance, midpoint, and slope for the following pairs of points:

12.1: $(1,2); (3,4)$ **12.2:** $(-5,2); (3,8)$

12.3: $(-1,10); (-2,6)$ **12.4:** $(1,0); (-1,2)$

12.5: $(-5,-5); (2,2)$ **12.6:** $(4,-3); (8-1)$

Find the slope-intercept equation for the line:

12.7: For Problem 12.1:
$(1,2); (3,4)$

12.8: For Problem 12.2:
$(-5,2); (3,8)$

12.9: For Problem 12.3:
$(-1,10); (-2,6)$

12.10: For Problem 12.4:
$(1,0); (-1,2)$

12.11: For Problem 12.5:
$(-5,-5); (2,2)$

12.12: For Problem 12.6:
$(4,-3); (8-1)$

12.13: $2y + 4x = 6$ **12.14:** $2(3 - y) = 6x$

12.15: $4 - y = 5x + 2$ **12.16:** $x + 5 = 9 - y$

Chapter 13

Solve by Combination:

13.1: $2y = 2x + 3; \ y = 4x$

13.2: $y = x - 2; \ 2y = 3x + 2$

13.3: $3y = x^2 + 8; \ y = 4 - x$

13.4: $4y = x^2 + 7; \ y = x + 1$

13.5: $y = x + 3; \ 3y = 2x^2 + 10$

13.6: $y = -x; \ 2y = x^2 - 3$

Solve by Substitution:

13.7: $y = x + 1; \ 4y = 5x - 1$

13.8: $y = 3x + 3; \ 2y = x + 10$

13.9: $y = 4x - 2; \ 3y = 4x^2 + 2$

13.10: $y = 10x + 4; \ \frac{y}{2} = 4x + 3$

13.11: $y = 2x^2; \ 3y = 5x + 1$

13.12: $y = x - 1; \ 2y = x^2 - 1$

Solve by Graphing Solutions:

13.13: $y = 3x - 2; \ 2y = 4x + 5$

13.14: $y = x + 2; \ y = x^2$

13.15: $y = 4x - 1; \ y = x^3 - 3$

13.16: $y = 3x + 5; \ y = x^2 + 1$

Chapter 14

14.1: Sandra is 16 years old which is twice as old as Sally was three years ago. How old is Sally today?

14.2: Lily got an excellent score on her last exam. Erica was jealous because she only got 75% of Lily's high score. Together they scored 168 points. What were the girls' scores?

14.3: It takes Jeff three times as long to walk his bike up a hill than to ride it down the hill. If the round trip takes 10 minutes, how long does is take Jeff to walk his bike up the hill?

14.4: Joe has three less than twice as many pairs of shoes as Jerry who has five pairs. How many pairs of shoes does Joe have?

14.5: Larry has one less than three times as many stamps as Moe. Curly has one more than six times as many stamps as Moe. Together the group has 100 stamps. How many does Moe have?

14.6: Sue sold 32 baseball cards for $40.00. Some sold for $1.55 each; the rest sold for $1.07. How many cards did she sell for $1.55?

14.7: Ayesha is training for a race and wants to consume 2000 calories a day. Within that allotment, she wants to consume twice as many calories from protein as carbohydrates, and she wants to consume 10% of her calories in fat. How many calories of protein will she consume?

14.8: A movie house sells 150 tickets for an event for a total of $3090.00. Adult ticket prices are $25.00 and child ticket prices are $13.00. How many adult tickets were sold?

14.9:   Ed, John, and Li divide up a pack of baseball cards so that Li takes one third, John takes one fourth of what remains, and Ed gets the 12 cards that are left. How many cards were in the pack?

14.10:   Nordstrom promises to buy one pack of cookies from the local girl scout every day for one week. Unfortunately, the price of the cookies goes up $1.00 every day. In the end, Nordstrom pays $56.00. What was the original price of the cookies?

14.11:   Jindal has two brothers. One brother is eight years older, and the other brother is two years younger. The combined ages of the boys is 36. How old is Jindal?

14.12:   Jackie has $50 to spend on dinner for her and her friends. For every three people she invites, the restaurant will charge a $4.00 service fee. If each dinner costs $7.00, how many can attend?

14.13:   Tickets for a community theater production have different prices for adults versus children. With an even split of adults to children, 100 tickets sold on Friday for $2200.00. On Saturday, there were an additional 25 adults and the sales were $2950.00. How much is the adult ticket?

14.14:   Curtis and Beatrice had a yard sale. Curtis sold three toys and two games for $4.25. Sue sold four toys and one game for $4.00. How much did they charge for the toys?

14.15:   If a typical slice of bread has 90 calories and 5 grams of protein, and a typical slice of cheese has 120 calories and 8 grams of protein, what combination would give 510 calories and 31 grams of protein?

14.16:   The cost of a long distance call to California is a nominal charge per call plus a certain rate per minute. If a 30 minute call costs $30, and a 20 minute call costs $21.00, how much is the rate per minute?

Appendix C: Solutions to Problem Sets

Chapter 1

1.1: 249
 	
Not even, so 2 and other even numbers are not a factors
Last digit not divisible by 5, so 5 and its multiples are not factors
2+4+9=15 which is divisible by 3, so 249 is divisible by 3:
     249/3=83
83 is not divisible by 3 (or its multiples), 7, 11,13,17
Factors are 3 and 83
	1.2: 512
	
	512 is even, so 2 is a factor which we'll apply until it's not even:
	        512
	  → 256 × 2
	  → 128 × 2
	  → 64 × 2
	  → 32 × 2
	  → 16 × 2
	  → 8 × 2
	  → 4 × 2
	  → 2 × 2
	All of the above are factors of 512

1.3: 1017
 	
Not even, so 2 and other even numbers are not a factors
Last digit not divisible by 5, so 5 and its multiples are not factors
1+0+1+7=9 which is divisible by 3:
     1017/3=339
     339/3=113
113 is not divisible by 3 (or its multiples), 7, 11,13,17
Factors are 3, 339, 113,and 9 (3x3)
	1.4: 52
	
	Divisible by 2:
	     52/2=26
	     26/2=13
	13 is prime
	Factors are 2, 26, 13, and 4 (2X2)

1.5: 100
 	
Divisible by 2:
     100/2=50
     50/2=25
     25/5=5
Factors are 2, 50, 25, 5, 4 (2X2), 10 (2X5), and 20 (4X5)
	1.6: 88
	
	Divisible by 2:
	     88/2=44
	     44/2=22
	     22/2=11
	Factors are 2, 44, 22, 11, 4 (2x2), and 8 (4X2)

1.7: 99 Not even, so not divisible by even numbers 9+9=18 which is divisible by 3: $\qquad$ 99/3=33 $\qquad$ 33/3=11 Factors are 3, 33, 11, and 9 (3X3)	**1.8:** 45 Not even, so not divisible by even numbers 4+5=9 which is divisible by 3: $\qquad$ 45/3=15 $\qquad$ 15/3=5 Factors are 3, 15, 5 and 9 (3X3)
1.9: $5 + (-9) - (-4)$ $\qquad = 5 - 9 + 4$ $\qquad = -4 + 4$ $\qquad = 0$	**1.10:** $2 + 8 - (-3)$ $\qquad = 2 + 8 + 3$ $\qquad = 10 + 3$ $\qquad = 13$
1.11: -2×8 $\qquad = -16$	**1.12:** $-3 \times -4 \times -2$ $\qquad = +12 \times -2$ $\qquad = -24$
1.13: $-14/(-7)$ $\qquad = +2$	**1.14:** $-14/|-7|$ $\qquad = -14/7$ $\qquad = -2$
1.15: $-|22|/|-2|$ $\qquad = -22/2$ $\qquad = -11$	**1.16:** $-10 - |-5|$ $\qquad = -10 - 5$ $\qquad -15$

Chapter 2

2.1: $\frac{52}{4}$	**2.2:** $\frac{12}{56}$
$$\frac{52}{4} = \frac{26}{2} = 13$$	$$\frac{12}{56} = \frac{6}{28} = \frac{3}{14}$$
2.3: $\frac{2}{\sqrt{2}}$	**2.4:** $\frac{100}{\sqrt{20}}$
$$\frac{2}{\sqrt{2}} = \frac{2}{\sqrt{2}} \times \frac{\sqrt{2}}{\sqrt{2}} = \frac{2\sqrt{2}}{2} = \sqrt{2}$$	$$\frac{100}{\sqrt{20}} = \frac{100}{\sqrt{20}} \times \frac{\sqrt{20}}{\sqrt{20}} = \frac{100\sqrt{20}}{20} = 5\sqrt{20}$$ $$= 5\sqrt{4 \times 5} = (5 \times 2)\sqrt{5} = 10\sqrt{5}$$
2.5: $\frac{300}{3\sqrt{10}}$	**2.6:** $\frac{24}{2\sqrt{8}}$
$$\frac{300}{3\sqrt{10}} = \frac{100}{\sqrt{10}} \times \frac{\sqrt{10}}{\sqrt{10}} = \frac{100\sqrt{10}}{10} = 10\sqrt{10}$$	$$\frac{24}{2\sqrt{8}} = \frac{12}{\sqrt{8}} \times \frac{\sqrt{8}}{\sqrt{8}} = \frac{12\sqrt{8}}{8} = \frac{3\sqrt{8}}{2}$$ $$= \frac{3\sqrt{4 \times 2}}{2} = \frac{(3 \times 2)\sqrt{2}}{2} = 3\sqrt{2}$$
2.7: $\frac{2}{3} + \frac{4}{5}$	**2.8:** $\frac{3}{2} + \frac{-5}{6}$
$$\frac{2}{3} + \frac{4}{5} = \frac{10}{15} + \frac{12}{15} = \frac{22}{15}$$	$$\frac{3}{2} + \frac{-5}{6} = \frac{9}{6} - \frac{5}{6} = \frac{4}{6} = \frac{2}{3}$$
2.9: $\frac{5}{12} - \frac{4}{20}$	**2.10:** $\frac{2}{3} - \frac{3}{4} + \frac{1}{6}$
$$\frac{5}{12} - \frac{4}{20} = \frac{25}{60} - \frac{12}{60} = \frac{13}{60}$$	$$\frac{2}{3} - \frac{3}{4} + \frac{1}{6} = \frac{8}{12} - \frac{9}{12} + \frac{2}{12}$$ $$= \frac{-1}{12} + \frac{2}{12} = \frac{1}{12}$$

2.11: $-\frac{3}{2} \times \frac{2}{5}$

$$-\frac{3}{2} \times \frac{2}{5} = -\frac{6}{10} = -\frac{3}{5}$$

2.12: $\frac{3}{4} \times \frac{1}{12} \times \frac{2}{3}$

$$\frac{3}{4} \times \frac{1}{12} \times \frac{2}{3} = \left(\frac{3}{48}\right) \times \frac{2}{3} = \frac{2}{48} = \frac{1}{24}$$

In second step, the 3's cancel

2.13: $\frac{1}{4} \div \frac{7}{8}$

$$\frac{1}{4} \div \frac{7}{8} = \frac{1}{4} \times \frac{8}{7} = \frac{8}{28} = \frac{2}{7}$$

2.14: $\frac{2}{6} \div \frac{-1}{7}$

$$\frac{2}{6} \div \frac{-1}{7} = \frac{2}{6} \times \frac{-7}{1} = -\frac{14}{6} = -\frac{7}{3}$$

2.15: 2500 in^3 to ft^3

We know there are 12inches in one foot, so we can set up the unit conversion:

$$\frac{2500 \text{ in} \times \text{in} \times \text{in}}{} \left| \frac{1 \text{ ft}}{12 \text{ in}} \right| \frac{1 \text{ ft}}{12 \text{ in}} \left| \frac{1 \text{ ft}}{12 \text{ in}} \right| = \frac{2500}{12^3} \text{ ft}^3$$

$$= 1.45 \text{ ft}^3$$

2.16: $20\frac{\text{feet}}{\text{second}}$ to $\frac{\text{miles}}{\text{hour}}$

There are 5280 feet in one mile, 60 seconds in a minute, and 60 minutes in an hour:

$$\frac{20 \text{ feet}}{\text{sec}} \left| \frac{1 \text{ mile}}{5280 \text{ feet}} \right| \frac{60 \text{ sec}}{1 \text{ min}} \left| \frac{60 \text{ min}}{1 \text{ hour}} \right| = \frac{20 \times 60 \times 60}{5280} \text{miles/hour}$$

$$= 13.6 \text{ mph}$$

Chapter 3

3.1: $5^2 \times 5$ $$= 5^2 \times 5^1 = 5^{2+1} = 5^3 = 125$$	**3.2:** $2^2 \times 2^4$ $$= 2^{2+4} = 2^6 = 64$$
3.3: $4^{-2} \times 4^2$ $$= 4^{-2+2} = 4^0 = 1$$	**3.4:** $4^2 \times 2^3$ $$= (2^2)^2 \times 2^3 = 2^{2\times2} \times 2^3 = 2^{4+3}$$ $$= 2^7 = 128$$
3.5: $8^3 \div 8^4$ $$= 8^{3-4} = 8^{-1} = \frac{1}{8}$$	**3.6:** $4^2 \div \sqrt{4}$ $$4^2 \div 4^{1/2} = 4^{2-1/2} = 4^{3/2}$$ $$= \sqrt{4^3} = \sqrt{64} = 8$$
3.7: $\frac{1}{7} \times 7^3$ $$= 7^{-1} \times 7^3 = 7^{-1+3} = 7^2 = 49$$	**3.8:** $4^{3/2} \times \frac{1}{2}$ $$= (2^2)^{3/2} \times 2^{-1} = 2^{6/2} \times 2^{-1}$$ $$= 2^{3-1} = 2^2 = 4$$
3.9: $5 \times (2 - 3)$ $$= 5 \times (-1) = -5$$	**3.10:** $3 \times \left(4 - (8 + 2)\right)$ $$= 3(4 - 10) = 3(-6) = -18$$

3.11: $(2^2 + 3)^2$ $(4 + 3)^2 = 7^2 = 49$	**3.12:** $\left(\frac{18}{(4-2)}\right)\left(\frac{(-2)^2}{6}\right)$ $= \left(\frac{18}{2}\right)\left(\frac{4}{6}\right) = \frac{3}{2}\left(\frac{4}{1}\right) = 6$
3.13: $\lvert(-2)^3(4-3)\rvert$ $= \lvert(-2)(-2)(-2)(1)\rvert = \lvert-8\rvert = 8$	**3.14:** $\sqrt{(4 \times 2^4)}$ $= (2^2 + 2^4)^{1/2} = (2^6)^{1/2} = 2^3 = 8$
3.15: $\sqrt[3]{5} \times \lvert-12 + 7\rvert^{-2}$ $= 5^{2/3} \times \lvert-5\rvert^{-2} = 5^{2/3} \times 5^{-2}$ $= 5^{\left(\frac{2}{3}-2\right)} = 5^{-4/6} = 5^{-2/3}$ $= \frac{1}{\sqrt[3]{5^2}} = \frac{1}{\sqrt[3]{25}}$	**3.16:** $\left((4-2)^{(1+2)} \times 8^2\right)^{-1/2}$ $(2^3 \times 8^2)^{-1/2} = (2^3 \times (2^3)^2)^{-1/2}$ $= (2^3 \times 2^6)^{-1/2} = (2^9)^{-1/2}$ $= 2^{-9/2} = 2^{-\left(4+\frac{1}{2}\right)}$ $= \frac{1}{2^{\left(4+\frac{1}{2}\right)}} = \frac{1}{2^4 \times 2^{1/2}} = \frac{1}{16\sqrt{2}}$

Chapter 4

4.1: $x + 5 = 2$ $x + 5 - 5 = 2 - 5$ $x = -3$	**4.2:** $x - 3 = 4$ $x - 3 + 3 = 4 + 3$ $x = 7$
4.3: $x + 9 = 10$ $x + 9 - 9 = 10 - 9$ $x = 1$	**4.4:** $x - 2 = 8$ $x - 2 + 2 = 8 - 2$ $x = 6$
4.5: $x + 2 = 13$ $x + 2 - 2 = 13 - 2$ $x = 11$	**4.6:** $x - 3 = 9$ $x - 3 + 3 = 9 - 3$ $x = 6$
4.7: $4x = 3$ $\frac{1}{4}(4x) = 3\left(\frac{1}{4}\right)$ $x = \frac{3}{4}$	**4.8:** $2x = 7$ $\frac{1}{2}(2x) = 7\left(\frac{1}{2}\right)$ $x = \frac{7}{2}$

4.9: $5x + 2 = -3$

$$5x + 2 - 2 = -3 - 2$$

$$5x = -5$$

$$\frac{1}{5}(5x) = -5\left(\frac{1}{5}\right)$$

$$x = -1$$

4.10: $4x + 5 = 9$

$$4x + 5 - 5 = 9 - 5$$

$$4x = 4$$

$$\frac{1}{4}(4x) = 4\left(\frac{1}{4}\right)$$

$$x = 1$$

4.11: $2x + 9 = 15$

$$2x + 9 - 9 = 15 - 9$$

$$2x = 6$$

$$\frac{1}{2}(2x) = 6\left(\frac{1}{2}\right)$$

$$x = 3$$

4.12: $3x - 2 = 7$

$$3x - 2 + 2 = 7 + 2$$

$$3x = 9$$

$$\frac{1}{3}(3x) = 9\left(\frac{1}{3}\right)$$

$$x = 3$$

4.13: $4x - 1 = 3$

$$4x - 1 + 1 = 3 + 1$$

$$4x = 4$$

$$x = 1$$

4.14: $2x + 3 = 7$

$$2x + 3 - 3 = 7 - 3$$

$$2x = 4$$

$$x = 2$$

4.15: $\frac{x}{5} + 3 = 5$

$$\frac{x}{5} + 3 - 3 = 5 - 3$$

$$\frac{x}{5} = 2$$

$$5\left(\frac{x}{5}\right) = 2(5)$$

$$x = 10$$

4.16: $\frac{2x}{3} - 1 = 3$

$$\frac{2x}{3} - 1 + 1 = 3 + 1$$

$$\frac{2x}{3} = 4$$

$$\frac{3}{2}\left(\frac{2x}{3}\right) = 4\left(\frac{3}{2}\right)$$

$$x = 6$$

Chapter 5

5.1: $4x^2 - 2 = 14$ $\qquad 4x^2 = 16$ $\qquad x^2 = 4$ $\qquad x = \pm 2$	**5.2:** $3x^2 - 3 = 0$ $\qquad 3x^2 = 3$ $\qquad x^2 = 1$ $\qquad x = \pm 1$
5.3: $\sqrt{x} - 4 = 0$ $\qquad \sqrt{x} = 4$ $\qquad \sqrt{x}^2 = 4^2$ $\qquad x = 16$	**5.4:** $3x^{-2} + \frac{1}{3} = \frac{2}{3}$ $\qquad \frac{3}{x^2} = \frac{2}{3} - \frac{1}{3} = \frac{1}{3}$ $\qquad \frac{1}{x^2} = \frac{1}{9}$ $\qquad x^2 = 9$ $\qquad x = \pm 3$
5.5: $x^3 + 4 = 12$ $\qquad x^3 = 8$ $\qquad \sqrt[3]{x^3} = \sqrt[3]{8}$ $\qquad x = 2$	**5.6:** $5x^2 + 5 = 50$ $\qquad 5x^2 = 45$ $\qquad x^2 = 9$ $\qquad x = \pm 3$
5.7: $4(x + 3) = 20$ $\qquad (x + 3) = 5$ $\qquad x = 2$	**5.8:** $2(x - 5) = -4$ $\qquad (x - 5) = -2$ $\qquad x = 3$

5.9: $\frac{x-2}{4} + 3 = 4$ $$\frac{x-2}{4} = 1$$ $$x - 2 = 4$$ $$x = 2$$	**5.10:** $3\left(\frac{x-2}{4}\right) - 6 = -6$ $$3\left(\frac{x-2}{4}\right) = 0$$ $$\frac{x-2}{4} = 0$$ $$x - 2 = 0$$ $$x = 2$$
5.11: $2(x^3 - 2) + 10 = 22$ $$2(x^3 - 2) = 12$$ $$(x^3 - 2) = 6$$ $$x^3 = 8$$ $$x = 2$$	**5.12:** $2(3x^2 - 30) + 6 = 42$ $$2(3x^2 - 30) = 36$$ $$(3x^2 - 30) = 18$$ $$3x^2 = 48$$ $$x^2 = 16$$ $$x = \pm 4$$
5.13: $4|x - 3| = 16$ $$|x - 3| = 4$$ $$x - 3 = \pm 16$$ Getting rid of the absolute value gives two equations: $x - 3 = 16 \qquad x - 3 = -16$ $x = 19 \qquad\quad x = -13$	**5.14:** $-2|x + 2| + 4 = 2$ $$-2|x + 2| = -2$$ $$x + 2 = \pm 1$$ Getting rid of the absolute value gives two equations: $x + 2 = 1 \qquad x + 2 = -1$ $x = -1 \qquad\quad x = -3$
5.15: $2(x^2 - 3) + 5 = 1$ $$2(x^2 - 3) = -4$$ $$(x^2 - 3) = -2$$ $$x^2 = 1$$ $$x = \pm 1$$	**5.16:** $8(x^{-1/3} + 2) - 20 = 4$ $$8(x^{-1/3} + 2) = 24$$ $$(x^{-1/3} + 2) = 3$$ $$\frac{1}{\sqrt[3]{x}} = 1$$ $$\sqrt[3]{x} = 1$$ $$\sqrt[3]{x}^3 = 1^3$$ $$x = 1$$

Chapter 6

6.1: $2x - 3x^2 = -4x^2 + 3x$ $\qquad -3x^2 + 4x^2 = 3x - 2x$ $\qquad x^2(-3 + 4) = x(3 - 2)$ $\qquad\qquad x^2 = x$ Note: $x = 0$ is one solution, so we can get rid of it by dividing through by x $\qquad\qquad x = 1$ Solutions are $x = 0$ and $x = 1$.	**6.2:** $x^2 - 3x = 2x - 4x^2$ $\qquad x^2 + 4x^2 = 2x + 3x$ $\qquad x^2(1 + 4) = x(2 + 3)$ $\qquad\qquad 5x^2 = 5x$ Note: $x = 0$ is one solution, so we can get rid of it by dividing through by $5x$ $\qquad\qquad x = 1$ Solutions are $x = 0$ and $x = 1$.
6.3: $3x + 8 = 4x - 2$ $\qquad 3x - 4x = -2 - 8$ $\qquad x(3 - 4) = -10$ $\qquad\qquad -x = -10$ $\qquad\qquad x = 10$	**6.4:** $5x - 2 = 8x + 1$ $\qquad 5x - 8x = 1 + 2$ $\qquad x(5 - 8) = 3$ $\qquad\qquad -3x = 3$ $\qquad\qquad x = -1$
6.5: $x = 2(x + 7)$ $\qquad x = 2x + 14$ $\qquad x - 2x = 14$ $\qquad x(1 - 2) = 14$ $\qquad\qquad -x = 14$ $\qquad\qquad x = -14$	**6.6:** $4(x + 2) = 2(x + 7)$ $\qquad 4x + 8 = 2x + 14$ $\qquad 4x - 2x = 14 - 8$ $\qquad x(4 - 2) = 6$ $\qquad\qquad 2x = 6$ $\qquad\qquad x = 3$
6.7: $4x = 3(x - 4)$ $\qquad 4x = 3x - 12$ $\qquad 4x - 3x = -12$ $\qquad x(4 - 3) = -12$ $\qquad\qquad x = -12$	**6.8:** $4 - x = 7(x + 3)$ $\qquad 4 - x = 7x + 21$ $\qquad x(-1 - 7) = 21 - 4$ $\qquad x(-8) = 17$ $\qquad\qquad x = -\dfrac{17}{8}$

6.9: $5(x^{-1} + 2) = \frac{3x+4}{x}$

$$\frac{5}{x} + 10 = 3 + \frac{4}{x}$$

$$\frac{5}{x} - \frac{4}{x} = 3 - 10$$

$$\frac{1}{x}(5 - 4) = -7$$

$$\frac{1}{x} = -7$$

$$x = -\frac{1}{7}$$

6.10: $\frac{x-2}{3x} = \frac{4}{x} + 3$

$$\frac{1}{3} - \frac{2}{3x} = \frac{4}{x} + 3$$

$$\frac{-2}{3x} - \frac{4}{x} = 3 - \frac{1}{3}$$

$$\frac{1}{x}\left(-\frac{2}{3} - 4\right) = \frac{8}{3}$$

$$-\frac{14}{3x} = \frac{8}{3}$$

$$\frac{1}{x} = \frac{-8}{14}$$

$$x = \frac{-14}{8} = \frac{-7}{4}$$

6.11: $5(\sqrt{x} + 2) = 3\sqrt{x} + 14$

$$5\sqrt{x} + 10 = 3\sqrt{x} + 14$$

$$5\sqrt{x} - 3\sqrt{x} = 14 - 10$$

$$2\sqrt{x} = 4$$

$$\sqrt{x} = 2$$

$$x = 4$$

6.12: $x^2 + 5 = 9(x^2 - 2)$

$$x^2 + 5 = 9x^2 - 18$$

$$x^2 - 9x^2 = -18 - 5$$

$$-8x^2 = -23$$

$$x^2 = 2.88$$

$$x = \pm 1.7$$

6.13: $2x^2 + 9 = 3x^2 + 3$

$$2x^2 + 9 = 3x^2 + 3$$

$$2x^2 - 3x^2 = 3 - 9$$

$$-x^2 = -6$$

$$x^2 = 6$$

$$x = \pm\sqrt{6}$$

6.14: $x\left(x - \frac{2}{x}\right) = 7(3 - x^2)$

$$x^2 - 2 = 21 - 7x^2$$

$$x^2 + 7x^2 = 21 + 2$$

$$8x^2 = 23$$

$$x^2 = 2.88$$

$$x = \pm 1.7$$

6.15: $x^3 - 1 = 3x(x^2 - \frac{2}{x})$

$$x^3 - 1 = 3x^3 - 6$$

$$-2x^3 = -5$$

$$x = \sqrt[3]{5/2}$$

6.16: $2\sqrt[3]{x} + 3 = 3x^{1/3} + 1$

$$2\sqrt[3]{x} + 3 = 3\sqrt[3]{x} + 1$$

$$-\sqrt[3]{x} = -2$$

$$x = 8$$

Chapter 7

7.1: A is the set of real numbers greater than 100. So, $x > 100$, and $x \in \mathbb{R}$: $A = \{x: x > 100; x \in \mathbb{R}\}$	**7.2:** A is the set of real numbers between 25 and 1000. $25 < x < 1000$, and $x \in \mathbb{R}$: $A = \{x: 25 < x < 1000; x \in \mathbb{R}\}$
7.3: A is the set of integers between 0 and 10. $x = \{1,2,3,4,5,6,7,8,9\}$ or $A = \{x: 0 < x < 10; x \in \mathbb{Z}\}$	**7.4:** A is the set of rational numbers greater than 2. $x > 2$, and $x \in \mathbb{Q}$: $A = \{x: x > 100; x \in \mathbb{Q}\}$
7.5: $A = \{x: x > 6; x \in B\}$ $\quad\ B = \{x: 0 \le x \le 10; x \in \mathbb{Z}\}$ We want to find all elements in B that are not in A: $A' = \{x: 0 \le x \le 6;\ x \in \mathbb{Z}\}$	**7.6:** $A = \{x: 3 < x < 7; x \in B\}$ $\quad\ B = \{x: 0 \le x \le 10; x \in \mathbb{Z}\}$ We want to find all elements in B that are not in A: $A' = \{x: 0 \le x \le 3; 7 \le x \le 10;\ x \in \mathbb{Z}\}$
7.7: $A = \{even\ integers; x \in B\}$ $\quad\ B = \{x: 0 \le x \le 10; x \in \mathbb{Z}\}$ A' would be the odd integers between 0 and 10: $A' = \{1,3,5,7,9\}$	**7.8:** $A = \{x: x > 5; x \in B\}$ $\quad\ B = \{x: 0 \le x \le 10; x \in \mathbb{R}\}$ A' would be numbers less than or equal to 5 and greater than or equal to 0: $A' = \{x: 0 \le x \le 5; x \in \mathbb{R}\}$
7.9: $A = \{x: x > 52; x \in B\}$ $\quad\ B = \{x: 0 \le x \le 100; x \in \mathbb{R}\}$ A' would be numbers less than or equal to 52 and greater than or equal to 0: $A' = \{x: 0 \le x \le 52; x \in \mathbb{R}\}$	**7.10:** $A = \{x: 25 < x < 50; x \in B\}$ $\quad\ B = \{x: 0 \le x \le 100; x \in \mathbb{R}\}$ A' has two intervals: $A' = \{x: 0 \le x \le 25; 50 \le x \le 100;$ $x \in \mathbb{R}\}$

7.11: $A = \{x : x > 6; x \in \mathbb{Z}\}$
$\quad\quad B = \{x : 0 \leq x \leq 10; x \in \mathbb{Z}\}$

$A \cup B$ is everything in A and B:

$$A \cup B = \{x : x \geq 0; x \in \mathbb{Z}\}$$

$A \cap B$ is everything in common between A and B:

$$A \cap B = \{x : 6 < x \leq 10; x \in \mathbb{Z}\}$$

7.12: $A = \{x : 3 < x < 7; x \in \mathbb{Z}\}$
$\quad\quad B = \{x : 0 \leq x \leq 10; x \in \mathbb{Z}\}$

$A \cup B$ is everything in A and B. Since $A \subseteq B$:

$$A \cup B = B = \{x : 0 \leq x \leq 10; x \in \mathbb{Z}\}$$

$A \cap B$ is everything in common between A and B:

$$A \cap B = A = \{x : 3 < x < 7; x \in \mathbb{Z}\}$$

7.13: $A = \{x : x > 6; x \in \mathbb{R}\}$
$\quad\quad B = \{x : 0 \leq x \leq 10; x \in \mathbb{Z}\}$

$$A \cup B = \{x : x = \{0,1,2,3,4,5\};$$
$$x \geq 6; x \in \mathbb{R}\}$$

$$A \cap B = \{x : 6 < x \leq 10; x \in \mathbb{Z}\}$$

7.14: $A = \{x : x < 0; x \in \mathbb{R}\}$
$\quad\quad B = \{x : x > 4; x \in \mathbb{R}\}$

$$A \cup B = \{x : x < 0; x > 4; x \in \mathbb{R}\}$$

$$A \cap B = \emptyset$$

7.15: $A = \{x : 0 < x < 25; x \in \mathbb{R}\}$
$\quad\quad B = \{x : 2 \leq x \leq 10; x \in \mathbb{R}\}$

$$A \cup B = A = \{x : 0 < x < 25; x \in \mathbb{R}\}$$

$$A \cap B = B = \{x : 2 \leq x \leq 10; x \in \mathbb{R}\}$$

7.16: $A = \{x : -5 \leq x \leq 5; x \in \mathbb{R}\}$
$\quad\quad B = \{x : x > 0; x \in \mathbb{R}\}$

$$A \cup B = \{x : x \geq -5; x \in \mathbb{R}\}$$

$$A \cap B = \{x : 0 < x \leq 5; x \in \mathbb{R}\}$$

Chapter 8

8.1: $2x + 5 > 2$ $$2x > -3$$ $$x > -\frac{3}{2}$$	**8.2:** $7x - 3 < 4$ $$7x < 7$$ $$x < 1$$
8.3: $-4x + 9 \geq 1$ $$-4x \geq -8$$ $$x \leq 2$$	**8.4:** $-2x - 2 < 8$ $$-2x < 10$$ $$x > -5$$
8.5: $-x + 2 \leq 13$ $$-x \leq 11$$ $$x \geq -11$$	**8.6:** $4x - 3 > 9$ $$4x > 12$$ $$x > 3$$
8.7: $-4(x + 2) < 8$ $$(x + 2) > -2$$ $$x > -4$$	**8.8:** $2(x - 8) < 10$ $$(x - 8) < 5$$ $$x < 13$$
8.9: $4x + 3 < 5x + 2$ $$4x - 5x < 2 - 3$$ $$-x < -1$$ $$x > 1$$	**8.10:** $5 - x \geq 9x - 5$ $$-x - 9x \geq -5 - 5$$ $$-10x \geq -10$$ $$x \leq 1$$

8.11: $\frac{2x}{3} + 1 > x - 5$

$$\frac{2x}{3} - x > -5 - 1$$

$$-\frac{x}{3} > -6$$

$$x < 18$$

8.12: $\frac{x}{5} - 2 \leq x - 3$

$$\frac{x}{5} - x \leq -3 + 2$$

$$\frac{-4x}{5} \leq -1$$

$$x \geq \frac{5}{4}$$

8.13: $-4x^2 + 5 < 3x^2 - 2$

$$-4x^2 - 3x^2 < -2 - 5$$

$$-7x^2 < -7$$

$$x^2 > 1$$

$$|x| > 1$$

$(x > 1$ and $x < -1$ as x^2 has two roots$)$

8.14: $-2(x^3 + 4) > 24$

$$(x^3 + 4) < 12$$

$$x^3 < 8$$

$$x < 2$$

8.15: $|x - 3| \geq 5$

We have two problems:

$x - 3 \geq 5$ $x - 3 \leq -5$

$x \geq 8$ $x \leq -2$

8.16: $|3 - 2x| < 7$

We have two problems:

$3 - 2x < 7$ $3 - 2x > -7$

$-2x < 4$ $-2x > -10$

$x > -2$ $x < 5$

Chapter 9

9.1: $(3x^2 + x + 4) - (x + 2)$	**9.2:** $(x^2 + 4x - 3) - (3x^2 - 4)$
$-(x + 2) = -x - 2$	$-(3x^2 - 4) = -3x^2 + 4$
$$\begin{array}{r} (3x^2 + x + 4) \\ +\ \ \ \ \ \ -x - 2 \\ \hline 3x^2 \ \ \ \ \ + 2 \end{array}$$	$$\begin{array}{r} (x^2 + 4x - 3) \\ +\ -3x^2 \ \ \ \ \ + 4 \\ \hline -2x^2 + 4x + 1 \end{array}$$
9.3: $(2x^2 + x - 2) - (x + 8)$	**9.4:** $(x^2 + 4x + 1) - (3x + 4)$
$-(x + 8) = -x - 8$	$-(3x + 4) = -3x - 4$
$$\begin{array}{r} (2x^2 + x - 2) \\ +\ \ \ \ \ \ -x - 8 \\ \hline 2x^2 \ \ \ \ \ - 10 \end{array}$$	$$\begin{array}{r} (x^2 + 4x + 1) \\ +\ \ \ \ \ \ -3x - 4 \\ \hline x^2 \ + x \ - 3 \end{array}$$
9.5: $(2x + 1)(x^2 - 2x + 1)$	**9.6:** $(3x - 3)(x^2 + 2x + 5)$
$$\begin{array}{r} (x^2 - 2x + 1) \\ \times \ \ \ \ \ \ 2x + 1 \\ \hline x^2 - 2x + 1 \\ 2x^3 - 4x^2 + 2x \ \ \ \ \ \\ \hline 2x^3 - 3x^2 \ \ \ \ \ + 1 \end{array}$$	$$\begin{array}{r} (x^2 + 2x + 5) \\ \times \ \ \ \ \ \ 3x - 3 \\ \hline -3x^2 - 6x - 15 \\ 3x^3 + 6x^2 + 15x \ \ \ \ \ \\ \hline 3x^3 + 3x^2 + 9x - 15 \end{array}$$
9.7: $(2x^2 - 3x + 9)(x^2 - x + 3)$	**9.8:** $(x^4 + x^3 + 1)(x^2 + x + 1)$
$$\begin{array}{r} (2x^2 - 3x + 9) \\ \times \ \ \ \ \ \ x^2 - x + 3 \\ \hline 6x^2 - 9x + 27 \\ -2x^3 + 3x^2 - 9x \ \ \ \ \ \\ 2x^4 - 3x^3 + 9x^2 \ \ \ \ \ \ \ \ \ \ \ \\ \hline 2x^4 - 5x^3 + 18x^2 - 18x + 27 \end{array}$$	$$\begin{array}{r} (x^4 + x^3 + 1) \\ \times \ \ \ \ \ \ x^2 + x + 1 \\ \hline x^4 + x^3 \ \ \ \ \ \ \ \ + 1 \\ x^5 + x^4 \ \ \ \ \ \ \ \ + x \\ x^6 + x^5 \ \ \ \ \ \ \ \ + x^2 \ \ \ \ \ \ \\ \hline x^6 + 2x^5 + 2x^4 + x^3 + x^2 + x + 1 \end{array}$$

9.9: $(2x + 1)(x - 1)$ $2x^2 - 2x + x - 1$ $= 2x^2 - x + 1$	**9.10:** $(x + 3)(x - 5)$ $x^2 - 5x + 3x - 15$ $= x^2 - 2x + 15$
9.11: $(-x - 2)(x + 1)$ $-x^2 - x - 2x - 2$ $= -x^2 - 3x - 2$	**9.12:** $(4x - 7)(2x + 3)$ $8x^2 + 12x - 14x - 21$ $= 8x^2 - 2x - 21$
9.13: $(3x + 1)(3x - 1)$ $9x^2 - 3x + 3x - 1 = 9x^2 - 1$	**9.14:** $(-2x - 4)(4x + 2)$ $-8x^2 - 4x - 16x - 8$ $= -8x^2 - 20x - 8$
9.15: $\left(\frac{x}{2} + 3\right)\left(\frac{x}{4} - 2\right)$ $\left(\frac{1}{2}\right)\left(\frac{1}{4}\right)x^2 - x + \frac{3}{4}x - 6$ $= \frac{x^2}{8} - \frac{x}{4} - 6$	**9.16:** $\left(\frac{2x}{3} - 1\right)\left(\frac{x}{3} + 1\right)$ $\left(\frac{2}{3}\right)\left(\frac{1}{3}\right)x^2 + \frac{2}{3}x - \frac{1}{3}x - 1$ $= \frac{2x^2}{9} + \frac{x}{3} - 1$

Chapter 10

10.1: $x^2 + 3x = x - 1$	**10.2:** $x^2 + 3x = -2$

10.1: $x^2 + 3x = x - 1$

$$x^2 + 2x + 1 = 0$$

1,1   1,1

$\boxed{+}$ $1(1) + 1(1) = 2$

$$(x + 1)(x + 1) = 0$$
$$x = \{-1\}$$

10.2: $x^2 + 3x = -2$

$$x^2 + 3x + 2 = 0$$

1,1   2,1

$\boxed{+}$ $1(2) + 1(1)$

$$(x + 2)(x + 1) = 0$$
$$x = \{-2, -1\}$$

10.3: $x^3 + 5x^2 = x^2 - 3x$

$$x(x^2 + 4x + 3) = 0$$

1,1   3,1

$\boxed{+}$ $1(3) + 1(1)$

$$x(x + 3)(x + 1) = 0$$
$$x = \{0, -3, -1\}$$

10.4: $2x^2 + 1 = -5x$

$$2x^2 + 5x + 1 = 0$$

Have to use the quadratic equation:

$$x = \frac{-5 \pm \sqrt{5^2 - 4(2)(1)}}{2(2)} = \frac{-5 \pm \sqrt{17}}{4}$$
$$x = \{-0.22, -2.3\}$$

10.5: $x^4 + x^2 = 7x^2 - 5$

$$x^4 - 6x^2 + 5 = 0$$

1,1   5,1

$\boxed{+}$ $1(5) + 1(1)$

$$(x^2 - 5)(x^2 - 1) = 0$$
$$x^2 = \{5, 1\}$$
$$x = \{\sqrt{5}, -\sqrt{5}, 1, -1\}$$

10.6: $x^2 = 4x - 4$

$$x^2 - 4x + 4 = 0$$

1,1   4,1 ; 2,2

$\boxed{+}$ $1(2) + 1(2)$

$$(x - 2)(x - 2) = 0$$
$$x = \{2\}$$

10.7: $x^3 = 5x^2 - 6x$

$$x(x^2 - 5x + 6) = 0$$

1,1   6,1 ; 2,3

$\boxed{+}$ $1(2) + 1(3)$

$$x(x - 2)(x - 3) = 0$$
$$x = \{0, 2, 3\}$$

10.8: $3x^2 = 11x - 10$

$$3x^2 - 11x + 10 = 0$$

3,1   1,10 ; 5,2

$\boxed{+}$ $1(5) + 3(2)$

$$(3x - 5)(x - 2) = 0$$
$$x = \{\tfrac{5}{3}, 2\}$$

10.9: $x^2 = 2 - x$

$$x^2 + x - 2 = 0$$

1,1   2,1

$1(2) - 1(1)$

$$(x + 2)(x - 1) = 0$$
$$x = \{-2, 1\}$$

10.10: $-x^2 = 2x - 3$

$$x^2 + 2x - 3 = 0$$

1,1   3,1

$1(3) - 1(1)$

$$(x + 3)(x - 1) = 0$$
$$x = \{-3, 1\}$$

10.11: $x^2 = 5 - 4x$

$$x^2 + 4x - 5 = 0$$

1,1   5,1

$1(5) - 1(1)$

$$(x + 5)(x - 1) = 0$$
$$x = \{-5, 1\}$$

10.12: $x^2 + \dfrac{7x}{2} = 2$

$$2x^2 + 7x - 4 = 0$$

1,2   2,2; 4,1

$2(4) - 1(1)$

$$(x + 4)(2x - 1) = 0$$
$$x = \left\{-4, \tfrac{1}{2}\right\}$$

10.13: $x^2 = x + 6$

$$x^2 - x - 6 = 0$$

1,1   6,1; 3,2

$1(3) - 1(2)$

$$(x - 3)(x + 2) = 0$$
$$x = \{3, -2\}$$

10.14: $x^2 = 1 + 2x$

$$x^2 - 2x - 1 = 0$$

Have to use the quadratic equation:

$$x = \frac{+2 \pm \sqrt{2^2 - 4(1)(-1)}}{2(1)} = \frac{2 \pm \sqrt{8}}{2} = 1 \pm \sqrt{2}$$
$$x = \{2.4, -0.4\}$$

10.15: $x^2 = 1$

$$x^2 - 0x - 1 = 0$$

1,1   1,1

$1(1) - 1(1)$

$$(x - 1)(x + 1) = 0$$
$$x = \{1, -1\}$$

10.16: $3x^2 = 7x + 6$

$$3x^2 - 7x - 6 = 0$$

1,3   6,1; 3,2

$3(3) - 1(2)$

$$(x - 3)(3x + 2) = 0$$
$$x = \left\{3, -\tfrac{2}{3}\right\}$$

Chapter 11

11.1: $2x - y = 5$

$y = 2x - 5$

X	Y
-4	-13
-3	-11
-2	-9
-1	-7
0	-5
1	-3
2	-1
3	1
4	3

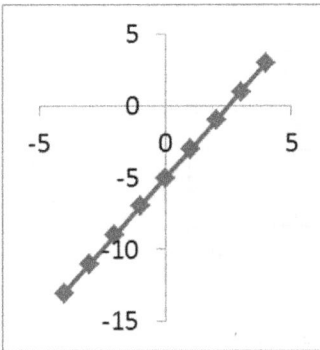

11.2: $4x + 2y = 12$

$y = 6 - 2x$

X	Y
-4	14
-3	12
-2	10
-1	8
0	6
1	4
2	2
3	0
4	-2

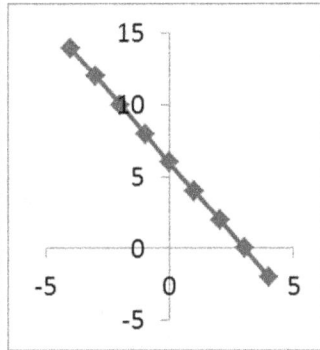

11.3: $y - 3x = 9$

$y = 3x + 9$

X	Y
-4	-3
-3	0
-2	3
-1	6
0	9
1	12
2	15
3	18
4	21

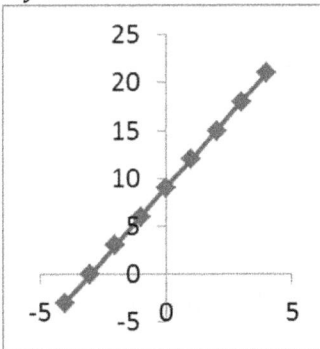

11.4: $y - 2x = 10$

$y = 2x + 10$

X	Y
-4	2
-3	4
-2	6
-1	8
0	10
1	12
2	14
3	16
4	18

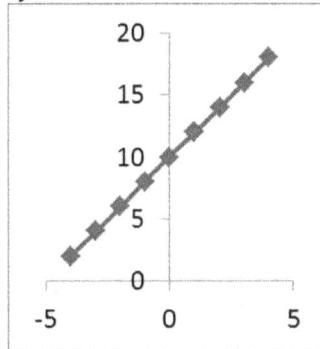

11.5: $y + 4x = 13$

$y = -4x + 13$

X	Y
-4	29
-3	25
-2	21
-1	17
0	13
1	9
2	5
3	1
4	-3

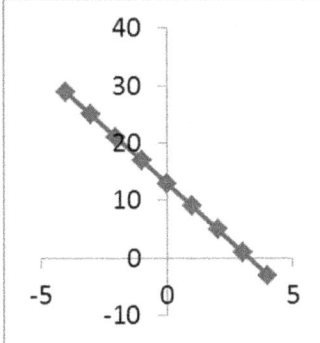

11.6: $4 - y = 3x$

$y = -3x + 4$

X	Y
-4	16
-3	13
-2	10
-1	7
0	4
1	1
2	-2
3	-5
4	-8

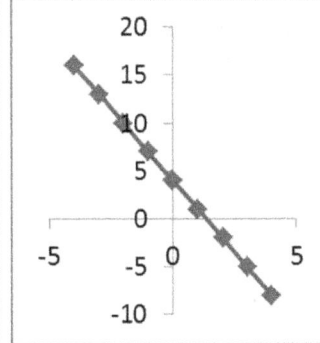

11.7: $y - x^2 = 2$
$$y = x^2 + 2$$

X	Y
-4	18
-3	11
-2	6
-1	3
0	2
1	3
2	6
3	11
4	18

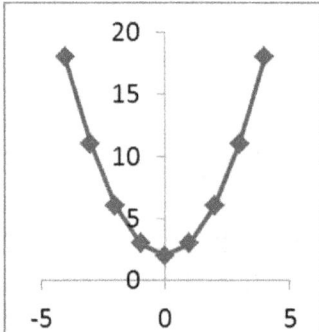

11.8: $y - 2x^2 + 3(x - 2) = 0$
$$y = 2x^2 - 3x + 6$$

X	Y
-4	50
-3	33
-2	20
-1	11
0	6
1	5
2	8
3	15
4	26

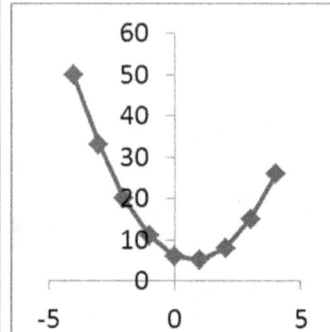

11.9: $x(x - 3) = y - x$
$$y = x^2 - 2x$$

X	Y
-4	24
-3	15
-2	8
-1	3
0	0
1	-1
2	0
3	3
4	8

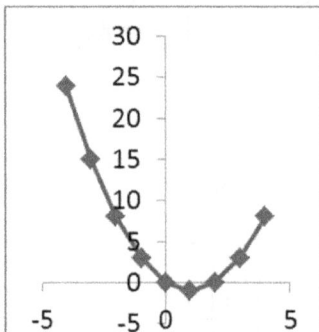

11.10: $4x + 2x^3 = 2y$
$$y = x^3 + 2x$$

X	Y
-4	-72
-3	-33
-2	-12
-1	-3
0	0
1	3
2	12
3	33
4	72

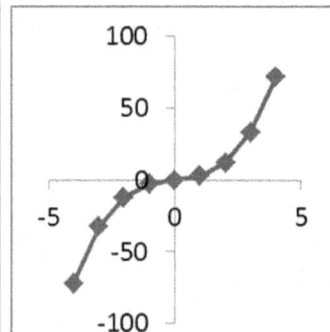

11.11: $(x - 2)^2 = y$
$$y = x^2 - 4x + 4$$

X	Y
-4	36
-3	25
-2	16
-1	9
0	4
1	1
2	0
3	1
4	4

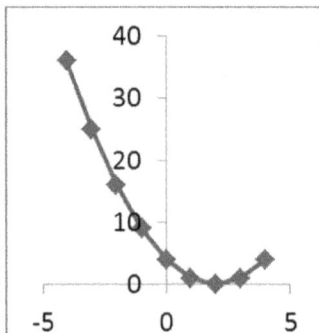

11.12: $x^4 - x^2 = y + 2$
$$y = x^4 - x^2 - 2$$

X	Y
-4	238
-3	70
-2	10
-1	-2
0	-2
1	-2
2	10
3	70
4	238

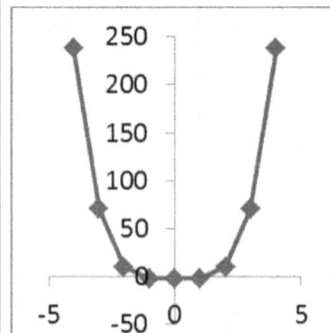

11.13: $z = 4x + 2y$

$z = 0 \rightarrow y = -2x$
$z = 1 \rightarrow y = \frac{1}{2} - 2x$
$z = 2 \rightarrow y = 1 - 2x$

Z=0		Z=1		Z=2	
X	Y	X	Y	X	Y
-4	8	-4	8.5	-4	9
-3	6	-3	6.5	-3	7
-2	4	-2	4.5	-2	5
-1	2	-1	2.5	-1	3
0	0	0	0.5	0	1
1	-2	1	-1.5	1	-1
2	-4	2	-3.5	2	-3
3	-6	3	-5.5	3	-5
4	-8	4	-7.5	4	-7

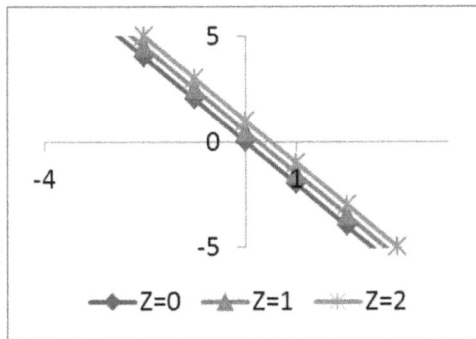

In 3-D, the plot looks like this:

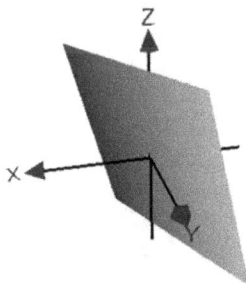

11.14: $z = y + x^2 + 2$

$z = 0 \rightarrow y = -2 - x^2$
$z = 1 \rightarrow y = -1 - x^2$
$z = 2 \rightarrow y = -x^2$

Z=0		Z=1		Z=2	
X	Y	X	Y	X	Y
-4	-18	-4	-17	-4	-16
-3	-11	-3	-10	-3	-9
-2	-6	-2	-5	-2	-4
-1	-3	-1	-2	-1	-1
0	-2	0	-1	0	0
1	-3	1	-2	1	-1
2	-6	2	-5	2	-4
3	-11	3	-10	3	-9
4	-18	4	-17	4	-16

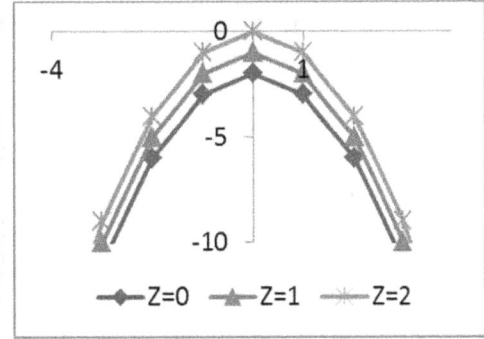

In 3-D, the plot looks like this:

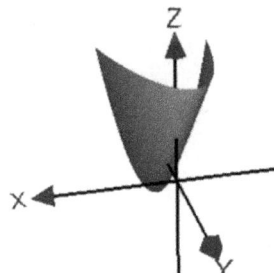

11.15: $z = y - x^3$

$z = 0 \rightarrow y = x^3$
$z = 1 \rightarrow y = 1 + x^3$
$z = 2 \rightarrow y = 2 + x^3$

Z=0		Z=1		Z=2	
X	Y	X	Y	X	Y
-4	-64	-4	-63	-4	-62
-3	-27	-3	-26	-3	-25
-2	-8	-2	-7	-2	-6
-1	-1	-1	0	-1	1
0	0	0	1	0	2
1	1	1	2	1	3
2	8	2	9	2	10
3	27	3	28	3	29
4	64	4	65	4	66

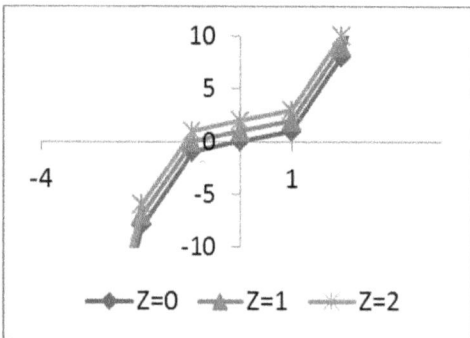

In 3-D, the plot looks like this:

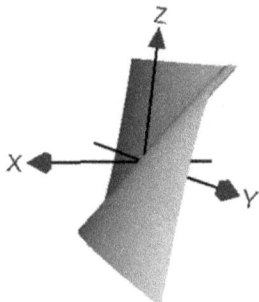

11.16: $z = y - 2x^2$

$z = 0 \rightarrow y = 2x^2$
$z = 1 \rightarrow y = 1 + 2x^2$
$z = 2 \rightarrow y = 2 + 2x^2$

Z=0		Z=1		Z=2	
X	Y	X	Y	X	Y
-4	32	-4	33	-4	34
-3	18	-3	19	-3	20
-2	8	-2	9	-2	10
-1	2	-1	3	-1	4
0	0	0	1	0	2
1	2	1	3	1	4
2	8	2	9	2	10
3	18	3	19	3	20
4	32	4	33	4	34

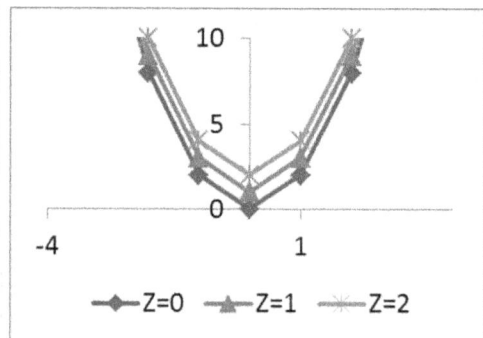

In 3-D, the plot looks like this:

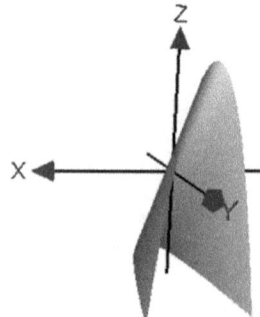

Chapter 12

12.1: $(1,2); (3,4)$ $d = \sqrt{(3-1)^2+(4-2)^2} = \sqrt{8} = 2\sqrt{2}$ $midpoint = \left(\frac{3+1}{2}, \frac{4+2}{2}\right) = (2,3)$ $m = \frac{4-2}{3-1} = 1$	**12.2:** $(-5,2); (3,8)$ $d = \sqrt{(3+5)^2+(8-2)^2} = 10$ $midpoint = \left(\frac{3-5}{2}, \frac{8+2}{2}\right) = (-1,5)$ $m = \frac{8-2}{3+5} = \frac{6}{8} = \frac{3}{4}$
12.3: $(-1,10); (-2,6)$ $d = \sqrt{(-2+1)^2+(6-10)^2} = \sqrt{17}$ $midpoint = \left(\frac{-2-1}{2}, \frac{6+10}{2}\right) = \left(-\frac{3}{2}, 8\right)$ $m = \frac{6-10}{-2+1} = \frac{-4}{-1} = 4$	**12.4:** $(1,0); (-1,2)$ $d = \sqrt{(-1-1)^2+(2-0)^2} = 2\sqrt{2}$ $midpoint = \left(\frac{-1+1}{2}, \frac{2+0}{2}\right) = (0,1)$ $m = \frac{2-0}{-1-1} = \frac{2}{-2} = -1$
12.5: $(-5,-5); (2,2)$ $d = \sqrt{(2+5)^2+(2+5)^2} = 7\sqrt{2}$ $midpoint = \left(\frac{2-5}{2}, \frac{2-5}{2}\right) = \left(-\frac{3}{2}, -\frac{3}{2}\right)$ $m = \frac{2+5}{2+5} = 1$	**12.6:** $(4,-3); (8-1)$ $d = \sqrt{(8-4)^2+(-1+3)^2} = 2\sqrt{5}$ $midpoint = \left(\frac{8+4}{2}, \frac{-1-3}{2}\right) = (6,-2)$ $m = \frac{-1+3}{8-4} = \frac{2}{4} = \frac{1}{2}$
12.7: For Problem 12.1: $\qquad (1,2); (3,4)$ Plug into point slope form with $(a,b) = (1,2)$ and $m = 1$, then reduce: $y - 2 = 1(x - 1)$ $y = x + 1$	**12.8:** For Problem 12.2: $\qquad (-5,2); (3,8)$ Plug into point slope form with $(a,b) = (-5,2)$ and $m = \frac{3}{4}$, then reduce: $y - 2 = \frac{3}{4}(x + 5)$ $y = \frac{3}{4}x + \frac{23}{4}$

12.9: For Problem 12.3: $$(-1,10); (-2,6)$$ Plug into point slope form with $(a, b) = (-1,10)$ and $m = 4$, then reduce: $$y - 10 = 4(x + 1)$$ $$y = 4x + 14$$	**12.10:** For Problem 12.4: $$(1,0); (-1,2)$$ Plug into point slope form with $(a, b) = (1,0)$ and $m = -1$, then reduce: $$y - 0 = -1(x - 1)$$ $$y = -x + 1$$
12.11: For Problem 12.5: $$(-5, -5); (2,2)$$ Plug into point slope form with $(a, b) = (-5, -5)$ and $m = 1$, then reduce: $$y + 5 = 1(x + 5)$$ $$y = x + 10$$	**12.12:** For Problem 12.6: $$(4, -3); (8 - 1)$$ Plug into point slope form with $(a, b) = (4, -3)$ and $m = \frac{1}{2}$, then reduce: $$y - 4 = \frac{1}{2}(x + 3)$$ $$y = \frac{1}{2}x + \frac{11}{2}$$
12.13: $2y + 4x = 6$ Put the equation into the proper form: $$2y = -4x + 6$$ $$y = -2x + 3$$	**12.14:** $2(3 - y) = 6x$ Put the equation into the proper form: $$-2y = 6x - 6$$ $$y = -3x + 3$$
12.15: $4 - y = 5x + 2$ Put the equation into the proper form: $$-y = 5x - 2$$ $$y = -5x + 2$$	**12.16:** $x + 5 = 9 - y$ Put the equation into the proper form: $$y = 9 - x - 5$$ $$y = -x + 4$$

Chapter 13

13.1: $2y = 2x + 3; \quad y = 4x$

$$-2(y = 4x) \rightarrow -2y = -8x$$

$$\begin{array}{r} 2y = 2x + 3 \\ + \ -2y = -8x \\ \hline 0 = -6x + 3 \end{array}$$

$$x = \frac{3}{6} = \frac{1}{2}$$

$$y = 4x = 2$$

Solution is $\left(\frac{1}{2}, 2\right)$

13.2: $y = x - 2; \quad 2y = 3x + 2$

$$-2(y = x - 2) \rightarrow -2y = -2x + 4$$

$$\begin{array}{r} 2y = 3x + 2 \\ + \ -2y = -2x + 4 \\ \hline 0 = \quad x + 6 \end{array}$$

$$x = -6$$

$$y = x - 2 = -6 - 2 = -8$$

Solution is $(-6, -8)$

13.3: $3y = x^2 + 8; \quad y = 4 - x$

$$-3(y = 4 - x) \rightarrow -3y = -12 + 3x$$

$$\begin{array}{r} 3y = x^2 \quad + 8 \\ + \ -3y = \quad 3x - 12 \\ \hline 0 = x^2 + 3x - 4 \end{array}$$

$$(x + 4)(x - 1) = 0$$
$$x = \{-4, 1\}$$

$$y = 4 - x = 4 + 4 = 8;$$
$$y = 4 - x = 4 - 1 = 3;$$

Solutions are $(-4, 8), (1, 3)$

13.4: $4y = x^2 + 7; \quad y = x + 1$

$$-4(y = x + 1) \rightarrow -4y = -4x - 4$$

$$\begin{array}{r} 4y = x^2 \quad + 7 \\ + \ -4y = \quad -4x - 4 \\ \hline 0 = x^2 - 4x + 3 \end{array}$$

$$(x - 3)(x - 1) = 0$$
$$x = \{3, 1\}$$

$$y = x + 1 = 3 + 1 = 4;$$
$$y = x + 1 = 1 + 1 = 2;$$

Solutions are $(3, 4), (1, 2)$

13.5: $y = x + 3; \quad 3y = 2x^2 + 10$

$$-3(y = x + 3) \rightarrow -3y = -3x - 9$$

$$\begin{array}{r} 3y = 2x^2 \quad + 10 \\ + \;\; -3y = \quad\quad -3x - 9 \\ \hline 0 = 2x^2 - 3x + 1 \end{array}$$

$$(2x - 1)(x - 1) = 0$$
$$x = \left\{\tfrac{1}{2}, 1\right\}$$

$$y = x + 3 = 3 + \tfrac{1}{2} = \tfrac{7}{2};$$
$$y = x + 3 = 1 + 3 = 4;$$

Solutions are $\left(\tfrac{1}{2}, \tfrac{7}{2}\right), (1,4)$

13.6: $y = -x; \quad 2y = x^2 - 3$

$$-2(y = -x) \rightarrow -2y = 2x$$

$$\begin{array}{r} 2y = x^2 \quad - 3 \\ + \;\; -2y = \quad\quad 2x \\ \hline 0 = x^2 + 2x - 3 \end{array}$$

$$(x + 3)(x - 1) = 0$$
$$x = \{-3, 1\}$$

$$y = -x = 3;$$
$$y = -x = -1;$$

Solutions are $(-3,3), (1,-1)$

13.7: $y = x + 1; \quad 4y = 5x - 1$

Plug 1st equation into 2nd:

$$4(x + 1) = 5x - 1$$
$$4x + 4 = 5x - 1$$
$$-x = -5$$
$$x = 5$$

$$y = x + 1 = 5 + 1 = 6$$

Solution is (5,6)

13.8: $y = 3x + 3; \quad 2y = x + 10$

Plug 1st equation into 2nd:

$$2(3x + 3) = x + 10$$
$$6x + 6 = x + 10$$
$$5x = 4$$
$$x = \tfrac{4}{5}$$

$$y = 3x + 3 = \tfrac{12}{5} + \tfrac{15}{5} = \tfrac{27}{5}$$

Solution is $\left(\tfrac{4}{5}, \tfrac{27}{5}\right)$

13.9: $y = 4x - 2;\ 3y = 4x^2 + 2$

Plug 1st equation into 2nd:

$$3(4x - 2) = 4x^2 + 2$$
$$12x - 6 = 4x^2 + 2$$
$$4x^2 - 12x + 8 = 0$$
$$(4x - 4)(x - 2) = 0$$
$$x = \{1, 2\}$$

$$y = 4x - 2 = 4 - 2 = 2$$
$$y = 4x - 2 = 8 - 2 = 6$$

Solutions are $(1,2), (2,6)$

13.10: $y = 10x + 4;\ \frac{y}{2} = 4x + 3$

Plug 1st equation into 2nd:

$$\tfrac{1}{2}(10x + 4) = 4x + 3$$
$$5x + 2 = 4x + 3$$
$$x = 1$$

$$y = 10x + 4 = 10 + 4 = 14$$

Solution is $(1,14)$

13.11: $y = 2x^2;\ 3y = 5x + 1$

Plug 1st equation into 2nd:

$$3(2x^2) = 5x + 1$$
$$6x^2 = 5x + 1$$
$$6x^2 - 5x - 1 = 0$$
$$(x - 1)(6x + 1) = 0$$
$$x = \{1, -\tfrac{1}{6}\}$$

$$y = 2x^2 = 2$$
$$y = 2x^2 = \frac{2}{36} = \frac{1}{18}$$

Solutions are $(1,2), (-\tfrac{1}{6}, \tfrac{1}{18})$

13.12: $y = x - 1;\ 2y = x^2 - 1$

Plug 1st equation into 2nd:

$$2(x - 1) = x^2 - 1$$
$$2x - 2 = x^2 - 1$$
$$x^2 - 2x + 1 = 0$$
$$(x - 1)(x - 1) = 0$$
$$x = 1$$

$$y = x - 1 = 1 - 1 = 0$$

Solution is $(1,0)$

13.13: $y = 3x - 2; \quad 2y = 4x + 5$

Plot $y = 3x - 2; \quad y = 2x + \frac{5}{2}$

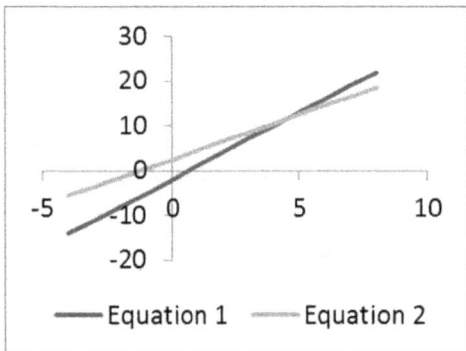

Solution is the crossover at $\left(\frac{9}{2}, \frac{23}{2}\right)$

13.14: $y = x + 2; \quad y = x^2$

Plot $y = x + 2; \quad y = x^2$

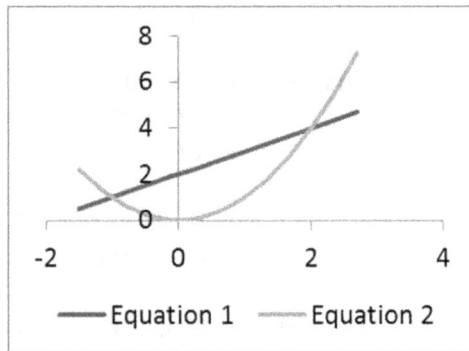

Solutions are the crossovers at $(-1,1), (2,4)$

13.15: $y = 4x - 1; \quad y = x^3 - 3$

Plot $y = 4x - 1; \quad y = x^3 - 3$

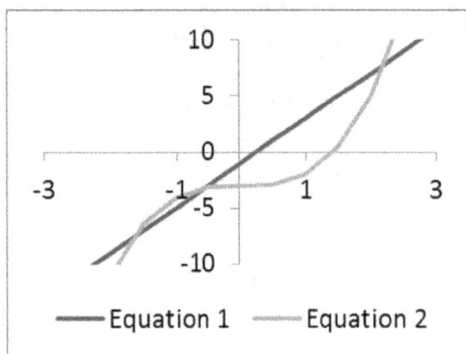

Solutions are the crossovers at approximately
$(-1.7, -7.7), (-0.5, -3.2), (2.2, 7.8)$

13.16: $y = 3x + 5; \quad y = x^2 + 1$

Plot $y = 3x + 5; \quad y = x^2 + 1$

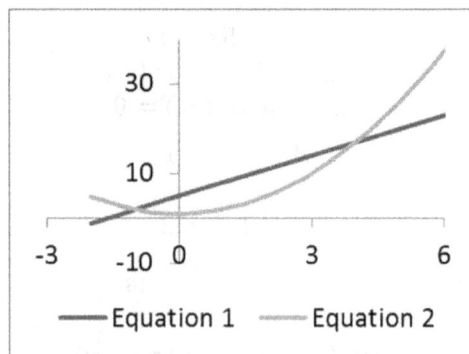

Solutions are the crossovers at
$(-1,2), (4,17)$

Chapter 14

14.1:	Sandra is 16 years old which is twice as old as Sally was three years ago. How old is Sally today? Sally is x years old. $$16 = 2(x - 3)$$ $$8 = x - 3$$ $$x = 11$$ Sally is 11 years old.
14.2:	Lily got an excellent score on her last exam. Erica was jealous because she only got 75% of Lily's high score. Together they scored 168 points. What were the girls' scores? Lily's score is x. Erica's score is $0.75x$. $$x + 0.75x = 168$$ $$x = \frac{168}{1.75} = 96$$ Lily scored 96, and Erica scored $0.75x = 72$.
14.3:	It takes Jeff three times as long to walk his bike up a hill than to ride it down the hill. If the round trip takes 10 minutes, how long does is take Jeff to walk his bike up the hill? Ride down time is x. Ride up time is 3x. $$x + 3x = 10$$ $$x = \frac{10}{4} = 2.5 \text{ min}$$ Ride up time is $3x = 7.5$ minutes.
14.4:	Joe has three less than twice as many pairs of shoes as Jerry who has five pairs. How many pairs of shoes does Joe have? Joe has $2x - 3$ pairs. Jerry has $x = 5$ pairs. Joe has $2(5) - 3$ pairs, or 7 pairs.

14.5: Larry has one less than three times as many stamps as Moe. Curly has one more than six times as many stamps as Moe. Together the group has 100 stamps. How many does Moe have?

Moe has x stamps, Larry has $3x - 1$, Curly has $6x + 1$.
Together they have 100:
$$x + 3x - 1 + 6x + 1 = 100$$
$$10x = 100$$
$$x = 10$$
Moe has 10 stamps.

14.6: Sue sold 32 baseball cards for $40.00. Some sold for $1.55 each; the rest sold for $1.07. How many cards did she sell for $1.55?

x cards sold for $1.55, $32 - x$ sold for $1.07.
$$1.55(x) + 1.07(32 - x) = 40$$
$$0.48x = 40 - 34.24 = 5.76$$
$$x = \frac{5.76}{0.48} = 12 \text{ cards sold for } \$1.55.$$

14.7: Ayesha is training for a race and wants to consume 2000 calories a day. Within that allotment, she wants to consume twice as many calories from protein as carbohydrates, and she wants to consume 10% of her calories in fat. How many calories of protein will she consume?

She will consume $0.10(2000) = 200$ calories from fat leaving 1800 calories divided between x carbs and $2x$ protein:
$$1800 = x + 2x$$
$$3x = 1800$$
$$x = 600$$

Ayesha will consume 1200 calories of protein.

14.8: A movie house sells 150 tickets for an event for a total of $3090.00. Adult ticket prices are $25.00 and child ticket prices are $13.00. How many adult tickets were sold?

$$25(x) + 13(150 - x) = 3090$$
$$12x = 3090 - 1950 = 250 = 1140$$
$$x = \frac{1140}{12} = 95 \text{ adult tickets.}$$

14.9:   Ed, John, and Li divide up a pack of baseball cards so that Li takes one third, John takes one fourth of what remains, and Ed gets the 12 cards that are left. How many cards were in the pack?

There are x cards.
Li gets $x/3$ cards, John takes $\frac{1}{4}(x - x/3)$, Ed gets 12.

$$x = \frac{x}{3} + \frac{1}{4}\left(\frac{2x}{3}\right) + 12$$

$$\frac{6x}{6} - \frac{2x}{6} - \frac{x}{6} = 12$$

$$\frac{3x}{6} = \frac{x}{2} = 12$$

$$x = 24 \text{ cards.}$$

14.10:   Nordstrom promises to buy one pack of cookies from the local girl scout every day for one week. Unfortunately, the price of the cookies goes up $1.00 every day. In the end, Nordstrom pays $56.00. What was the original price of the cookies?

Cookies cost x on day 1, on day 2 they cost $x + 1$, on day 3 they cost $x + 2$, etc.
$$7x + 0 + 1 + 2 + 3 + 4 + 5 + 6 = 56$$
$$7x = 56 - 21 = 35$$
$$x = 5$$

The cookies cost $5.00 on the first day.

14.11:   Jindal has two brothers. One brother is eight years older, and the other brother is two years younger. The combined ages of the boys is 36. How old is Jindal?

Jindal is x years old, his older brother is $x + 8$, and his younger brother is $x - 2$.
$$x + x + 8 + x - 2 = 36$$
$$3x = 36 - 6 = 30$$
$$x = 10 \text{ years old.}$$

14.12:	Jackie has $50 to spend on dinner for her and her friends. For every three people she invites, the restaurant will charge a $4.00 service fee. If each dinner costs $7.00, how many can attend? The service fee is 4/3 per person, and dinner is 7/person; $$\frac{4x}{3} + 7x = 50$$ $$\frac{25x}{3} = 50$$ $$x = \frac{50}{25}(3) = 6 \text{ people.}$$
14.13:	Tickets for a community theater production have different prices for adults versus children. With an even split of adults to children, 100 tickets sold on Friday for $2200.00. On Saturday, there were an additional 25 adults and the sales were $2950.00. How much is the adult ticket? 50 adults and 50 children attend Friday. 75 adults and 50 children attend Saturday. Cost for adults is x, and cost for children is y. $$50x + 50y = 2200$$ $$75x + 50y = 2950$$ Substitute $50y = 2200 - 50x$ into 2nd equation: $$75x + 2200 - 50x = 2950$$ $$25x = 2950 - 2200 = 750$$ $$x = 30$$ Cost for an adult ticket is $30.
14.14:	Curtis and Beatrice had a yard sale. Curtis sold three toys and two games for $4.25. Sue sold four toys and one game for $4.00. How much did they charge for the toys? Toys cost x, and games cost y. $$3x + 2y = 4.25$$ $$4x + y = 4$$ Substitute $y = 4 - 4x$ into 1st equation: $$3x + 8 - 8y = 4.25$$ $$-5x = -3.75$$ $$x = 0.75$$ The toys cost $0.75 each.

14.15: If a typical slice of bread has 90 calories and 5 grams of protein, and a typical slice of cheese has 120 calories and 8 grams of protein, what combination would give 510 calories and 31 grams of protein?

There will be x slices of bread and y slices of cheese.

$$90x + 120y = 510$$

$$5x + 8y = 31$$

$$y = \frac{31}{8} - \frac{5}{8}x$$

$$90x + 120\left(\frac{31}{8} - \frac{5}{8}x\right) = 510$$

$$90x + 465 - 75x = 510$$

$$15x = 45$$

$$x = 3$$

$$5x + 8y = 31$$

$$8y = 31 - 5(3) = 16$$

$$y = 2$$

There are 3 slices of bread and 2 slices of cheese to successfully make the combination.

14.16: The cost of a long distance call to California is a nominal charge per call plus a certain rate per minute. If a 30 minute call costs $30, and a 20 minute call costs $21.00, how much is the rate per minute?

The charge per call is x, and the fee per minute is y.

$$x + 30y = 30$$

$$x + 20y = 21$$

Substitute $x = 21 - 20y$ into the 1st equation.

$$21 - 20y + 30y = 30$$

$$10y = 9$$

$$y = 0.9$$

The fee per minute is $0.90.

Index

Other Books In This Series

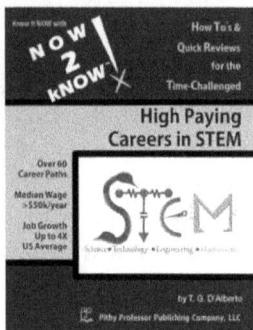

NOW 2 kNOW™ High Paying Careers in STEM

Plan for a career where your education dollars and hard work really pay off. STEM careers (science, technology, engineering, and math) are high paying, and companies are scrambling to find qualified candidates in the U.S. Better yet, the work is rewarding with opportunities to really impact the world!

NOW 2 kNOW™ Geometry

Geometry is the study of shapes and their parts. It is heavy in definitions, theorems, and proofs. The NOW 2 kNOW™ Geometry text organizes the subject's vast amount of information and provides numerous examples and practice with proofs!

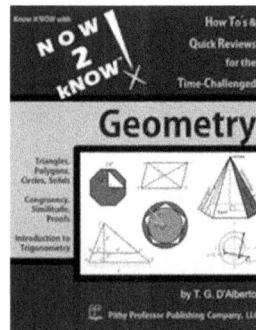

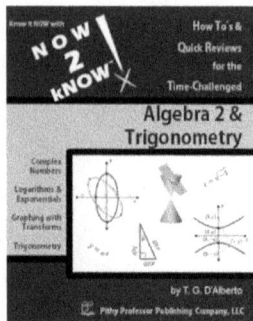

NOW 2 kNOW™ Algebra 2 & Trigonometry

Expand on the concepts of Algebra I and Geometry with this two course text! Thorough and concise instruction coupled with over 200 problems and worked out solutions will have you on your way to Calculus in no time!

NOW 2 kNOW™ Calculus I

Calculus is the gateway to many financially stable and successful careers. You might think this would make it hard and inaccessible, but that simply isn't true. Calculus is actually very easy! Once you see the concepts outlined succinctly, you'll see how little it takes to become a master.

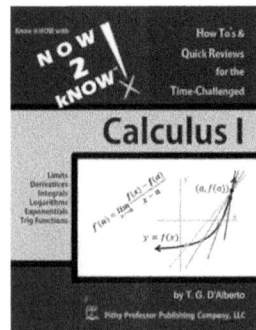

Go to Amazon.com and Search "NOW 2 kNOW"
or visit www.NOW2kNOW.com for updates.

www.ingramcontent.com/pod-product-compliance
Lightning Source LLC
Chambersburg PA
CBHW062027210326
41519CB00060B/7193